Irrigation and Irrigation Projects in India:
Tribunals, Disputes and Water Wars Perspective

Irrigation and Irrigation Projects in India:
Tribunals, Disputes and Water Wars Perspective

S. Jeevananda Reddy

Hyderabad.

BSP **BS Publications**

A unit of **BSP Books Pvt. Ltd.**

4-4-309/316, Giriraj Lane, Sultan Bazar,
Hyderabad - 500 095
Phone : 040 - 23445605, 23445688

Irrigation and Irrigation Projects in India:
Tribunals, Disputes and Water Wars Perspective by Dr. S. Jeevananda Reddy

© 2017, *by Publisher*

Published by :

BSP **BS Publications**
A unit of **BSP Books Pvt. Ltd.**

4-4-309/316, Giriraj Lane, Sultan Bazar,
Hyderabad - 500 095
Phone : 040 - 23445605, 23445688
e-mail : info@bspbooks.net

ISBN: 978-93-5230-187-4 (HB)

"In any irrigation projects in India, under interstate or intrastate, *might is right* play a pivotal role. This is the case with the Krishna River Water tribunal and Polavaram project"

"Don't move into Theoretical syndrome, it leads nowhere but drove into Practical Path gives you a right solution"

Preface

Science has empowered humankind with the knowledge and secrets of universe and the role human beings can play towards ensuring sustainable Planet, the "Earth" and ensuring a healthy and happy life within one's life span and transferring the pro planet growth systems and strategies to the future generation. Playing with the nature is like scratching the head with the fire. Yet we need development to meet the needs of non-linearly increasing population under the new lifestyles. But this development must meet the basic needs of people today without ruining the chances of future generations to do the same that comes under sustainable development. Though, now UN, Pope Francis, US President and Indian Government talk at the same wavelength but the actions proposed goes against sustainable development. This system wants to benefit few who use maximum resources at the cost of the remaining.

After the economic liberalization in 1992, the fundamental goal of successive governments has been economic growth for few at the cost of environmental degradation. India is among the 10 most industrialized countries in the world with the 8th largest economy. However, rapid economic and industrial growth is causing severe agriculture, urban and industrial pollution. India, thus, has been ranked as the most hazardous country in the world. This may create a situation for class war to meet their respective minimum needs in terms of natural resources, particularly "the Water", in coming decades.

Water is a natural renewable resource, fundamental to life, livelihood, food security, sustainable development. It is also the scarce resource. Thus water management plays vital role. We continued to overlook environmental damages until polluted land, water, food and air began to threaten human health and until native species & ecosystem began disappear. Resulting groundwater contamination, rivers become efficient channel to carry pollution, reservoirs became cesspools of poison, acid rain destroying pristine forests, water bodies & land. Developed countries instead of taking action on this global phenomenon, masked this scenario and brought the concept of global warming. Thus controlling of pollution, that has maximum impact on nature, has taken the back seat and global warming with little or no impact on nature took front seat as in this billions of dollars are there to share. To build the story, the author has taken some information from the internet. The author herewith Acknowledge the same with Thanks.

-Author

Recommendations

"Injustice anywhere is a threat to justice everywhere"

India being an agrarian state, in the agriculture production and thus improving the economy of the farmers and the states, irrigation and irrigation projects played major role for hundreds of years. Immediately after Independence, the first Prime Minister, Late Pandit Jawaharlal Nehru gave importance to irrigation sector wherein he considered 'Dams are Modern Temples'. Yet, around 60% of the cultivated area is at the mercy of "Rain God". This is basically because of the changes that took place in Indian political set up. The regional political parties ruled the roots of the states and nation. They neglected irrigation and agriculture resulting rapid increase in farmers' suicides and hunger and at the same time destruction of water resources. To reduce the disparity among dry-land and wet-land farming and thus improve the economy and nutritional security, there is a need to reduce the dependency on direct rainfall. This is achieved through irrigation and irrigation projects. There are large numbers of books that present the theoretical aspects but there are no books in practical perspective. The present book looked into the practical aspects pertaining to irrigation and irrigation projects in India. Such an information play vital role in future projects implementation including appointment of tribunals and understanding the disputes in right perspective and that help interlinking of rivers.

The delayed initiations of irrigation projects became cost-ineffective and the project related activities are taking decades to complete, for example Polavaram Project under Godavari River, though proposed in 1941, Godavari River Water Tribunal cleared the project in 70-80s, got national project status in 2009/2014 and finally Polavaram authority was created is still not moving as needed. The government says we will complete it by 2018. To achieve this goal, government must spend at least one-third of the total cost each year but allocating peanuts and thus it will take decades to complete. AP Cabinet on 11th October 2015 decided to raise the project cost to Rs. 36,000 crores from the 2011 cost of Rs. 16,000 crores. In 2005 it was Rs. 10,271 crores.

In this system the basic hitch is submergence and rehabilitation. The issue of submergence and rehabilitation is marred by politics. When political parties butcher thousands of acres of forest land to develop vote-bank settlements, nobody bothered but when the forest area submerges with an irrigation project hell will broke. When downstream villages are submerged due to floods every now and then nobody bothered but with a project when villages in upstream submerge hell will broke!!! Why this? Can't we streamline these? This must be done in the case of projects with public interest like irrigation projects. Forest departments take years to give clearance to lay drinking water pipeline through forests that escalates the cost as well delay the completion of the project; but least perturbed when Private investors butcher even mangroves in sensitive zones? Why this???

CWC must look into the issue of 'what is the affect of floods on downstream villages in natural scenario' before talking the issue of flood water from the dam in the upstream areas. In such cases, has government taking any action on upstream states for not building projects to use the allocated water? Affected and displaced estimations must be completed before the initiation of the projects and should not change the figures at any cost under R & R packages even under Political-bureaucratic-social Activists pressure. Most of them have their vested interests and thus create delays in the project execution. This is a major issue that hurdles the progress of the projects.

There is stiff opposition from certain quarters on dams and interlinking of rivers with arguments with little substance, sometimes to serve the vested interests. On a long term, India needs interlinking of rivers as even after 68 years of Independence more than 60% of the cultivated area is at the mercy of "Rain God". Rainfall is highly variable with space and time under changing climate. This is essential to build pollution free agriculture. The laws are weak and ambiguous. Indian water-dispute settlement mechanisms are opaque. Though water disputes are sometimes settled, with the new political culture with demand for water increasing, these lead to anarchy. This needs correction.

Though water balance data is very important, there is no centralized data bank to date though emphasized the need of such data in several recommendations. This must be achieved. Though several recommendations emphasized the need to take into account the climate in such studies, this was not given due place. Designing or re-designing projects without proper assessment of water availability will lead to anarchy and thus water wars. This is seen with Telangana State government that put forth new proposals every other day.

When governments change at state and central levels, first they want to redesign everything so that they get benefits irrespective of negative repercussions on environment, people and government. In the case of Krishna and Godavari Rivers, the water availability estimates at Dhavaleswaram and Prakasham barrages during the cyclonic activities, all the flood waters reaching these barrages are not available for use as it has to be released into the Sea after crossing the specified level at barrages. This is rarely taken into account while calculating the available water. This problem will be there to all such rivers along the east and west coasts. The CWC must look into this aspect and expect to rectify this. In this scenario, the government must come forward to change certain guidelines to minimize political-bureaucratic-activists interference on projects relating to irrigation.

Climate change is the major component of climate. Changing climate does not mean global warming which is insignificant to effect nature. We experienced the changes in the past for which our forefathers adapted technologies to overcome such variations; and in future also we encounter such variations in climate that needs adaptation to overcome the impact of vagaries of monsoons in India as they are highly regional specific and linked to general circulation patterns. Unfortunately government of India is not looking in this direction and betting hundreds of billions of dollars on emission scenarios. If the rainfall presents a natural variability – cyclic variation --, truncated data presents a misleading inferences such as increasing trend or decreasing trend or higher or lower values. To get meaningful results, the data must be of normal distribution; and skewed distribution will give biased conclusions. At least one cycle data is required to get meaningful inferences.

Surplus and deficits waters are either shared or lower riparian state may be allowed to suffer during deficit and benefit during surplus. Only surplus distribution is bad policy. This must be corrected. In India, projects are built without any clearances or basis. This type of activities creates anarchy and thus leads to water wars among states and among districts in a state. This is an excellent food for regional politicians. To avoid such scenarios, the Central Government must establish a mechanism to monitor and stop them until they are cleared by the appropriate authority. Unfortunately, activists present more bias over politicians in such cases. The Central Government must bring out an Act that limits the powers of politicians to meddle with the ongoing projects by either State or Central Governments to serve their vested interests. Otherwise these acts will become burden to public and the fruits of development will be delayed. The cost of completing the projects will double or treble with the delays. Classical example to this is Narmada and Polavaram projects.

Tribunals are filled with retired judges with political bent with unfettered powers and as a result they invariably bulldozed opposition presenting highly biased awards. Here the technical/scientific issues are taking back seat and legal issues are taking front seat. With this the beneficiaries are the legal luminaries and losers are the people of India. They put forth unscientific and illogical qualitative arguments to justify their actions. Also, through mathematical manipulations favour one state over the others. This is exactly what has happened with Krishna River Tribunal and unfortunately they also got extension of the term. Firstly the tribunal raised the water availability with the data manipulation and thus this affected the probability levels at which this water is available. Through mathematical manipulation the tribunal tried to show that even with increasing the dam height to store additional water still AP will be getting more than allocated water. Here, the tribunal has no respect for the predecessor tribunal award similar to a High Court Bench dumping a Supreme Court judgment into the dustbin on the same issue and ruling in its own way. To avoid such catastrophic awards, the Central Government should enact a Law to replace the tribunals with a technical body headed by a sitting Supreme Court Judge and a member from CWC will be the member convener; assisted by a 15 member team from different Central Government Departments. All will have a fixed two year term – they may hire temporary experts to resolve specific issues. In this connection, the government must formulate guidelines "clearly" so that nobody uses his or her vested interests into that to vitiate the justice. Very simple and less controversial award was delivered by Justice Bachawat and very complicated highly controversial and biased award was delivered by Justice Brijesh Kumar. To avoid such variations, the basics must be clearly defined without giving a scope to tribunal to manipulate to serve the vested interests.

Though the riparian states agreed and even signed for Polavaram project, with the change of guard, the present governments are opposing the very same project and creating hurdles in the progress of the project. Even the Central Government is not far behind those states and in fact played spoil sport to meet their political and regional interests. This type of political hurdles must be stopped. Therefore, to avoid such hurdles similar to above, the Central Government must establish a technical body to deal with inter-state and inter-region projects to avoid disputes and thus water wars to meet the political goals of ruling junta. AP government initiated Pattiseema lift irrigation project without any clearances to serve their vested needs by putting aside the main ongoing Polavaram project. Central government should not encourage such projects and put a foot on irresponsible projects like Pattiseema anywhere in the country. There is other issue in Polavaram that created big lobby to oppose the project as through Smuggling, Ganja Cultivation, Tourism and Mining thousands of crores are earned each year. Once the project is ready, they lose this lucrative tax free illegal income. We rarely look into this.

The Central and State Governments must develop a system to utilize the wastewater generated in urban areas before they are dumped into water bodies or rivers. Violators should be punishable on two counts – fine and jail terms. In the case of lift irrigation schemes including wells and bore-wells, government must make it mandatory to utilize them effectively using micro-irrigation practices and through the selection of less water intensive crops by discouraging the suicidal crops with high input costs along with this making crop rotation a mandatory. Permission to polluting activities must be kept outside the catchment areas of water resources.

About the Author

Dr. Sazzala Jeevananda Reddy is an Agrometeorologist got post-graduation in Geophysics & Applied Statistics with the advanced training in Meteorology & Oceanography and numerical weather prediction. Dr. Reddy got his Ph.D. in Agricultural Meteorology from the "The Australian National University", Canberra. Dr. Reddy has a wide experience in the field of Agrometeorology and Agroclimatology while working in several National and International institutions/organizations within and outside India. Dr. Reddy served Food and Agriculture Organization [FAO] as Expert & World Meteorological Organization [WMO] as Chief Technical Advisor.

The author, Dr. Sazzala Jeevananda Reddy, is one of the few scientists who started his carrier in the science of Climate Change as back as early 1970s. One of Dr. Reddy's work was identified in 1976 as one of the fifteen papers of unusual interest (most of them related to the subject of climate change) by SCOSTEP (Scientific Committee on Solar Terrestrial Physics) of American Academy of Sciences from among the entire literature published in National and International journals in Solar Terrestrial Physics up to that time. Dr. Reddy carried out analysis using observed meteorological data series over different parts of the globe. Using findings from such studies, Dr. Reddy presented concept of adopting agriculture in long-term agriculture planning in the semi-arid tropics. In this direction Dr. Reddy brought out several books to educate World Community; and published articles in National and International journals and as well presented at several National and International conferences. Reddy also contributed popular articles to daily news papers and magazines. The few important books contributed by Dr. Reddy are:

Reddy, S.J., 1993: 'Agroclimatic/Agrometeorological Techniques: As applicable to Dry-land Agriculture in developing countries', www.scribd.com, & Google Books, 205p – book review appeared in Agric. For. Meteorol., 67:325-327 [1994].

Gupta, R.K. & Reddy, S.J. (eds.), 1999: 'Advanced Technologies in Meteorology', 549p, Tata McGraw-Hill Publ. Comp. Ltd., New Delhi, India.

Reddy,   S.J.,   2002:   'Dry-land   Agriculture   of   India: An Agroclimatological and Agrometeorological perspective', 429p, BS Publ., Hyderabad, India

Reddy, S.J. 2008 & 2010: 'Climate Change: Myths & Realities', 176p & 114p, www.scribd.com & Google Books.

Reddy, S.J. 2011: '"Green" Green Revolution: Agriculture in the perspective of Climate Change", 160 p, www.scribd.com & Google Books.

Dr. Sazzala Jeevananda Reddy
Plot No. 277, Jubilee Hills, Phase-III;
Road No. 78, near Padmalaya Studio,
Hyderabad -500 096; Tel. (040) 23550480/23540762
E-Mail: jeevanandareddy@yahoo.com;
jeevananda_reddy@yahoo.com

Contents

Chapter 1: Introduction 1

Chapter 2: Water Resources Availability 13

Chapter 3: Inter-state Irrigation Projects – Role of Tribunals 43

List of Tables and Figures

List of Tables

List of Figures

Introduction

General: Water is a natural renewable resource, fundamental to life, livelihood, food & nutritional security, sustainable development. It is also a scarce resource. Two thirds of the Earth is covered by water, of which 97% of it is saline. That means fresh water covers only around 3%. However, of this 68.7% is in icecaps & glaciers and 30.1% as groundwater, wherein large part of it is not available for use as fresh water. Only small part of groundwater & ice melt is available as fresh water. Around 0.3% is only available as surface water and of which 87% is in lakes, 11% in swamps and 2% in rivers.

India has more than 17.11% of the world's population, but has only 4.6% of world's water resources with 2.3% of world's land area. The natural input to any surface water system is precipitation and snow melt within its watershed. In India, around 78% of average rainfall occurs in June to September, known as Southwest Monsoon. However, at regional level they are highly variable with orographic patterns – Western Ghats, Himalayan ranges, etc, Northeast Monsoon [October to December], Western disturbances in winter, cyclonic activity in pre-and post-monsoon seasons in Arabian Sea and Bay of Bengal. Humans have no control on precipitation & snowfall. They present, thus, high variations with space and time. Also, climate change [**Reddy, 2008 & 2010**] plays significant role in defining the rainfall and snowfall [**Reddy, 1993 & 2002**] with the time. All these define water availability in space and time.

Climate Change versus Global warming: UN introduced a new body to look in to this issue, namely IPCC [Inter-governmental Panel on Climate Change] a political body in place of WMO [World

Meteorological Organization] wherein meteorological services of different countries are part and parcel in developing policies relating to weather and climate. IPCC is filled with picked-up people to serve the vested interests. Though they talk of climate change but the studies primarily related to global warming and emissions. Also, global warming was built on false foundations, namely global temperature data series. This data series were built with meager data covering around 20% to 25% of the global area under variable met network distribution and ecological – land use and land cover – changes. This is manipulated time and again to meet their policy. Based on such data IPCC postulated several sensational hypotheses.

Global warming proponents knew the fact that global warming is not a settled science. Because of this global warming proponents instead say "climate change is settled Science", which is a fact. The main component of climate change is natural variability [**WMO, 1966**]. Humans have no control over it and thus we need to adapt to them – this I presented for agriculture in 70s & 80s and summarized in **Reddy [1993]**. They are region specific. Extremes are part of natural variability. This can be seen from the climate normal data of individual countries wherein extremes have not crossed these limits. Also, regional general circulation patterns over different seasons play the important role in year to year variations in extremes under natural variability. For example Western Disturbances in northwestern parts of India will influence heat and cold waves in summer and winter based on the high pressure belt location [**Reddy & Rao, 1978**]. The second component of climate change is ecological changes – associated with changes in land & water use and land cover changes. They are highly local-region specific. Urban-heat-island effect and rural-cold-island effect comes under this. They will influence the natural variability in that local/region. This plays important role in power consumption and agriculture.

The third is greenhouse effect on temperature. This is a natural change. Global warming is part of it but it is not settled science basically because: in nature the conversion component of natural CO_2 in to temperature has reached a saturation point. Any new addition to CO_2 due to man's actions, the increase in temperature is insignificant – a plateau like scenario [see Figure 1 in **Reddy (1995)**]. Because of this, IPCC with 97% scientific support [as they claim], IPCC using trial and error approach to link CO_2 raise to temperature raise under global warming. As a part of this game, IPCC goes on changing the "sensitivity factor – 1.95 under AR$ was reduced to 1.55 under AR5" that relates CO_2 with temperature. Also, IPCC now states global warming is more than half of the global temperature rise [50.1% is also more than half]. This is a subjective statement. That means that they agreed that the global temperature rise not entirely due to global warming phenomenon. IPCC has no valid data to verify its estimates as the global temperature curve is built on hypothetically derived data series as stated above. When that data series showed a hiatus for the last 19 years, three different theories were put forward – one group manipulates the data to remove hiatus, another group says temperature is hiding in deep-ocean and the third says it is a part of natural variation.   But all these never cared to look in to temperature curve built using balloon and satellite data. According to this using IPCC inference, global warming component is just around 0.15 °C, which is insignificant to influence climate or nature. Politicians don't understand these issues and politicians and scientists look at 100 billon dollars – Paris summit!!! Because of this, all those groups are using the word climate change instead of global warming.

Our leaders say something, talk something and act something else. There is no coherence in their speeches and to real goals on which they talk of. For example they quote sufferings of one 'X' and build the story. Finally they are falling in to the trap of global warming and CO_2. How this is going to save that 'X' -- hunger-starvation --? In the past activists brought out an album on 'Save Africa' collected billions of dollars and really the amount went in to that component is not even 10% and the rest went in to the pockets of the activists.

IT groups wanted to spread its roots in all aspects of life. This is highly power consuming activity. Instead of putting a cap on such activities, power production cannot be reduced and thus CO_2. Indian PM's visit in September 2015 to Silicon Valley, IT top bosses say that they will dump their IT into rural India, which in turn is going to effecting severely agriculture sector and health of education system. During previous PM time they wanted to spend few thousand crores on providing free cell phones to rural labour. I got it stopped on the same basis. Agriculture is the backbone to mitigate the goals UN announced and that is what US President saying and Pope Francis saying. The incoherent actions will lead nowhere!!!

We talk of global warming and climate change but we rarely look in to local environmental problems and resolving them appropriately. Take the example of festivals. They turn in to money spinning monsters along with vote bank provider. Governments through them are trying to divert peoples' attention on their failures. The impact of the festivals on environment and human health are enormous. Media talk and talk but indirectly canvassing in favour of those activities. This year's Ganesh Chaturdi in Hyderabad and its surroundings run in to around Rs. 20,000 crores business activities. In this used thousands of tons of plaster of Paris. This not only filling the lakes but also polluting the lakes -- this is indirectly serving the needs of realtors. Some of it removed and dumped in to low lying areas that making the area of no use and reducing infiltration of rain water in to ground, which is increasing year after year.

The chemical colours used in the Ganesha's making are polluting the water bodies and thus groundwater. Every year this is increasing. The festival organizers used to tell that this pollution is negligible when compared to regular pollution of Hussainsagar Lake in the heart of Hyderabad. The observations made by pollution control board, now, indicated nearly double in parameters after the immersion of Vinayaka's in the Lake. This increase is associated with discharging large part of the water in the lake in to Musi River and rainfall contribution was not much. In addition to this, the processions on the last day increased the air and noise pollution that severely affected the health of young kids and old. With all these negative effects, neither the government nor the people are interested to bring down the number of idols and the height of the idols – Khairatabad idol height is 60 feet. Unfortunately, in this case even the courts playing ding-dong role favouring the government. To resolve the issue, government appointed a single retired judge committee in 1985 and it recommended idols height should be below 3 feet [traditional Hindu system they are supposed to be 8" x 4". But, neither the courts nor the government interested to implement this recommendation. Media goes on saying groundwater is polluted and air pollution effecting the people, etc., every other day to get ratings. All this is lip-sympathy. Some Media propaganda says that they can build more than 60 feet idols with clay. This is the attitude!!! What difference it makes either we use clay or plaster of Paris in building the idols of that tall. Both fill the water bodies and reduce the water holding capacity.

Then next festival comes with crackers. It increases the pollution levels by more than 1000 times and cause health and physical harm but we do little to stop them. While manufacturing and storage accidents several children died and yet neither the government nor the people do their little bit to stop them!!! We are unable to stop them basically because: now a day festivals turned in to lucrative business running in to thousands of crores. Politicians-bureaucrats get their share. In fact the government itself is encouraging the same type in IT. This is in fact running in to lakhs of crores of business and at the same time harming the environment and human health. Large part of the power consumption is used in IT related industry. Finally the casualty is water and human health. Can we do something on this menace to protect ourselves in specific and people in general???

These are practical issues that can be easily stopped but we rarely interested but talk of global warming and emissions.

Water Management: Water management is the activity of planning, developing, distributing and managing the optimum uses of water resources. In an ideal world, water management planning has regarded to all the competing demands for water and seeks to allocate water on an equitable basis to satisfy all uses and demands. This is rarely possible in practice as under corrupt and bad governance the developmental activities rarely reach their goals. The projects run as per the dictates of judiciary, bureaucracy and politicians even if projects are technical in nature. Unfortunately technically qualified people including engineers & scientists rarely question such decisions or discuss in open – worried about their promotions and monitory gains and awards-rewards. Thus, technical issues are put in the hands of non-technicians to serve the political-region interests at the cost of people. In the river water distribution among riparian states, tribunals appointed by Central Government plays vital role. The tribunals are filled with retired judges with unfettered powers even though irrigation projects are technical in nature and less of legal in nature. The integrity of the judges plays an important role to get unbiased recommendations.

Sources of Irrigation: Among the different sources of irrigation except canal irrigation systems, others are localized with little conflicts in sharing of waters in the past; but it is not so any more. Regional political parties created war like conditions under vote bank politics and at the same time for monitory gains. The first Prime Minister of India, Late Pundit Jawaharlal Nehru observed that "dams are the modern temples" and with this spirit encouraged building of dams. This helped India to achieve self-sufficiency in food production. This helped India to reach from begging bowl scenario to exporting scenario. However, even after 68 years of Independence to India, the progress in irrigation projects are moving three steps forward and six steps backward with the changing political priorities of elected ruling parties and inter-state & within the regions in a state politics. The progress in irrigation projects severely hampered, thus, with regional political parties, whose priorities are looting the state under the disguise of populist schemes. Also regional political parties found irrigation is a major hurdle in their vote bank politics and thus still to date more than 60% of the cultivated area in India is at the mercy of "Rain God". To add to the vows, western funded NGOs became major hurdle in the progress of irrigation projects – particularly dams and inter-linking of rivers.

Agrarian Basis: Civilizations have been dependent on development of irrigated agriculture to provide agrarian basis of a society and to enhance the security of people. Globally irrigated area has been increased from 8 Mha in1800 to just over 255 Mha in 1995. With the passing of time the % area under different sources of irrigation changed drastically [**Reddy, 2014**]. The over exploitation of groundwater not only reduced the water quality but also reduced drastically the area cultivated per pump [it reduced from 2.5 ha to 0.5 ha in Andhra Pradesh], and thus resulted increase in power consumption per hectare. Also with the reduction in area under tanks indirectly reduced the capacity of recharging of groundwater.

In October 2015 FAO released a draft on "Sustainable agricultural development for food security and nutrition including the role of livestock". On this they released my comments on line on 5th October 2015, in which I observed that: I wonder whether the material presented in the document is meant to serve the interests of "IPCC" or to serve the people/nations to develop sustainable agriculture. From this report it is clear that the report is not really made to serve the people/nations to build sustainable agriculture.

Emissions and global warming has insignificant influence on agriculture. IPCC goes on changing the sensitivity factor that relates anthropogenic greenhouse gases with global warming. Also, now they say in a qualitative manner that it is only more than half of global temperature constitutes global warming. From the balloon and satellite data so far the global warming is only 0.15 °C. This is insignificant to influence agriculture. The seasonal and annual and dry to wet periods temperature variation goes beyond 10 °C that the local crops experiencing for the centuries.

The main component of climate change that has direct impact is natural variability. This I worked for several countries and adapted agriculture to them. Sometimes these are modified by local/regional ecological changes – land use and land cover changes. Based on this concept, forefathers developed agriculture systems that are sustainable under variable rainfall conditions. To improve the economy and nutrient security they adapted animal husbandry in to agriculture. This system was killed by chemical input mono crop agriculture. Here the yields increased with the level of irrigation and fertilizer supply and reached a plateau by 80s. So, the increase in production was due to chemical inputs. But, technology is not the primary cause for increasing the production but the primary component is irrigation. Agriculture under small holdings will be sustainable not with the technology but by providing supplemental water over rainfall.

The chemical input agricultural technology introduced the evil pollution [soil, water, air and food] that is affecting the health of life forms including the humans and crop production as well quality of potable water. To better utilize the scarce natural resources under small farm holdings is cooperative farming under organic inputs. These are the issues that the report should have concentrated rather than emissions and global warming to serve IPCC. I am more of a practical man than theoretical like IPCC which only creates panic and help collect billions of dollars to collect and share. Some of these are exposed very recently.

As part of 67th Independent Day address to the Nation our Respected President of India Shri. Pranab Mukherjee put fine words of "inclusive growth" but what is happening is selective growth. Under globalization scenario 25% of the population of the world including India consumes 75% of its natural resources annually. Unfortunately, world over sustainable growth or development in itself is not a political agenda but it has been recognized that without political change, sustainable growth or development is not possible. For example the present NDA government mandated projects line Digital India & Swacch Bharat running in to lakhs of crores but mostly benefitting the urban rich.

Production: Competition for land & water increased gradually with the growth in population, urbanization & industrialization. Though food production can meet the food needs of the population of India at present – but this is not meeting farmers' economic & nutrition security under chemical input technology, as it is not equally distributed resulting hunger & starvation. Though farmers developed pollution free innovations with organic inputs have the yield advantage, government is encouraging GMO's with no yield advantage but work under chemical input pollution and irrigation condition only [**Reddy, 2011**].

I think technology might have originated from man's laziness. He didn't want to use his muscle power like other animals to earn a living. So he used his intelligence to invent enhancements. Eventually he found that the enhancements did not really reduce his work, it only created a new set of problems which increased his workload. So he invented another technology to solve those problems. But again that new technology gave rise to a new set of problems and he invented another technology to solve those. The process is going on and on; we call it progress.

When we want adopt a new technology, we rarely look at its' long term impacts, both positive and negative, of such technologies on nature and thus on environment as it is driven by profit mode. This lacuna is glaringly evident in agriculture. In all these, business interests out play the environmental consequences. Sometimes it may not be possible to recover the destruction caused by the technologies, particularly those that affect biodiversity. Here the public relations groups play the spoil sport that mosque the reality.

Environmental concerns: Sometimes we equate environmental concerns of projects such as mining-industry to irrigation projects. The former is to benefit individual investors and the later to benefit people in general. We talk of cost effectiveness of the later but not the former. However, with the passing of time the fresh water in lakes and rivers are turning in to cesspools of poison [**Figure 1.1**]. The potable water availability for drinking is severely affected, though polluted water is still used in agriculture. The classical example is the water from the Musi River in Hyderabad – it is the same all around urban areas in the country. The water is used to cultivate around 2.5 lakh acres. Thus the food produced from such water is highly contaminated and creating numerous health hazards. To provide potable water huge sums are invested to treat the water for drinking. For cleaning Ganga River UPA spent around Rs. 20000 crores and now NDA government wanted to spend Rs. 1,00,000 crores with no end for getting pollution free Ganga water. Here we are not following the "precautionary principle".

Figure 1.1: River pollution

Pollution: Rachel Carson published "Silent Spring" in 1962. The book documented the detrimental effects on the environment—particularly on birds—of the indiscriminate use of pesticides. Carson accused the chemical industry of spreading disinformation and public officials of accepting industry claims unquestioningly. This has lead environmental movement in 70s. However, with powerful lobby this was thwarted and in its place a new concept was brought in. Billions of dollars were spent on this to control research & dissemination of false information. This is termed as global warming.

Pollution can take many forms. The air we breathe, the water we drink, the ground where we grow our food, and even the increasing noise we hear every day—all contribute to health problems and a lower quality of life. Water is the casualty of such pollution in several ways. This reduces the water availability for use under different sectors.

Judiciary: Though the judiciary is one of the four pillars of the Indian Constitution, it has become no different from other three pillars with reference to corruption. The 2014 reports in media high lights how our judiciary is functioning starting with their recruitment. Here the major issue is even after retirement they enjoy power. In fact, I think, this issue started with my letter to the Chief Justice of India with a copy to the Prime Minister of India on 11th February 2013. In this I raised three issues namely "not before me", "quid pro co" and "recruitment – collegiums system". Former two are illogical and are used to manipulate justice. In Andhra Pradesh to destroy the political rivals or to protect the criminals/corrupt politicians these two clauses were/are used invariably. The misuses of these two clauses are more hazardous than corruption. Justice is rarely achieved with such a system. Unfortunately in the recruitment of judges there is no independent body like UPSC and thus lacking integrity. Majority of the judges belong to particular industrial/business/political groups as they are behind their appointments/ recruitments. My letter was forwarded [may be] to the law ministry and law ministry organized a meeting with CMs, judges along with PM. CMs favoured UPSC but judges disagreed. However, government brought out National Judicial Appointment Commission [NJAC] Act, 2014 in place of existing collegiums system of recruitment and this was notified on April 15, 2015. Judges are fighting in Supreme Court on this for their power but when I submitted to three successive Chief Justices of the Supreme Court on fraudulent act of tribunal of Krishna Water sharing, they did not even cared to acknowledge it. -- Former Supreme Court judge justice Katju exposed the wrong doings by Chief Justices of India itself. A report says "SC favours I-T probe against former CJI's kin: Brought benami properties worth Crores" – 15th September 2015. With such a scenario with the judiciary, what will happen if such judges after retirement are appointed as River Water Tribunal members & Chairmen?

The other side of it is political interferences to gain vote bank at the cost of development and achieve easy monitory gains. With the regional parties, this scenario has aggravated. Completion of irrigation projects are taking decades. With the change of guard at the State and or at the Centre the priorities are changing. This not only is affecting the finances but also the expected development. In India we don't have clear cut mandates in irrigation projects, more particularly on inter-state projects. With this, real estate takes the priority over peoples' interest/needs.

In light of these, in this book, the author looked in to the aspects pertaining to irrigation and irrigation projects in India in general and Andhra Pradesh in specific to get sustainable development.

Water Resources Availability

2.1 Introduction

India being an agrarian country, water availability is an important input in to the system. Although the only natural input to any surface water system is precipitation and snow melt within the watershed, the total quantity of water in that system at any given time is also dependent on several factors. These factors include storage capacity in lakes, wetlands and artificial reservoirs, the permeability of the soil beneath these storage bodies, the runoff characteristics of the land in the watershed, the timing of the precipitation, soil condition and local evaporation rates. Also these factors also affect the proportions of water loss. Human activities can have large and sometime devastating impacts on these factors. Humans can increase storage capacity by constructing reservoirs and decrease it by draining wetlands. Humans often increase runoff quantities and velocities by paving areas and channelizing streams flow. Humans also influence the quality of available water through pollution factor. This makes the water in rivers and storages is affected severely and making it unsuitable-unsustainable for use. In this chapter discussed these issues in brief in general.

2.2 River Systems in India in Brief

Rivers are natural flowing water bodies, generally of fresh water, that flow towards either an ocean, a lake, a sea or another river. They are part of the hydrological cycle and the water in the rivers comes from different sources. They begin as small streams and gradually expand in size as more water gets added to them. The subcontinent of India has many rivers. The **Figure 2.1** presents the top 10 rivers in India and the length of these ten rivers given in Kilometers as: Indus – 2900, Brahmaputra – 2900, Ganga – 2510,

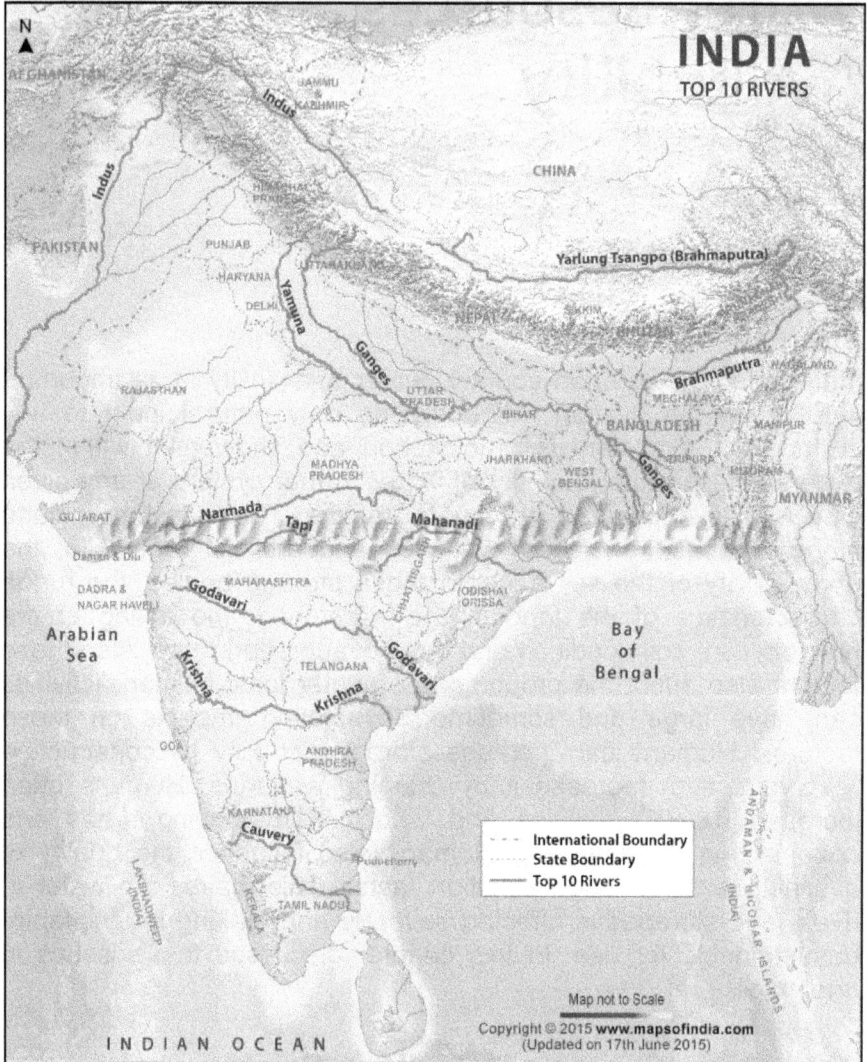

Figure 2.1: Principal rivers of India

Godavari – 1450, Narmada – 1290, Krishna – 1290, Yamuna – 1211, Mahanandi – 890, Kaveri – 760 and Tapi – 724. Nearly all the major cities of India are located on the banks of rivers. However, seven major rivers along with their numerous tributaries make up the river system of India. The largest basin systems of the rivers join the Bay of Bengal; however, some of the rivers whose courses take them through the western part of the country and towards the east of the state of Himachal Pradesh join the Arabian Sea. Parts of Ladek, northern parts of the Aravalli range and the arid parts of the Thar Desert have inland drainage. All major rivers of India originate from one of the three main wetlands, namely the Himalaya and the Karakoram ranges, Vindya and Satpura ranges and Chotanagpur plateau in central India, and Sahyadri or Western Ghats in western India.

The Indo-Gangetic Plains: This area is drained by 16 major rivers. The major Himalayan Rivers are the Indus, the Ganges, and the Brahmaputra. These rivers are long, and are joined by many large and important tributaries. Himalayan rivers have long courses from their source to sea. The Ganga and the Bramhaputra Rivers join Bay of Bengal. The renewable water source comes from rainfall and snow melt;

Indus River System: The Indus River originates in the northern slopes of the Kailas range near Lake Manasarovar in Tibet. Although most of the river's course runs through neighbouring Pakistan, a portion of it does run through Indian Territory, as do parts of the courses of its five major tributaries. These tributaries are the source of the name of the Punjab; the name is derived from the punch ("five") and aab ("water"), hence the combination of the words (Punjab) means "land with the water of five rivers". The renewable water source comes from rainfall and snow melt;

The Peninsular River System: The main water divide in peninsular rivers is formed by the Western Ghats, which run from north to south close to the western coast. Most of the major rivers of the Peninsula such as the Mahanadi, the Godavari, the Krishna and the Kaveri flow eastwards and drain into the Bay of Bengal. These rivers make delta at their mouths. The Narmada, Periyar and Tapti are the only long rivers, which flow west and make estuaries. The main source of renewable water is rainfall.

Majority of the rivers are originating in Hills/forests and finally joining the Arabian Sea, Bay of Bengal and Indian Ocean.

Orography plays vital role on the rainfall occurrences. In the case of Western Ghats on the windward side [Westside] get copious rains and the leeward side [east side], we call this as shadow zone, highly variable and poor rainfall occurs. Himalayan mountain ranges act as box, gives copious rains in northeast India – Chirapunji in Assam receives the highest rainfall in the world.

2.3 Water Resources Availability: Pollution

The main sources of water are precipitation and snow melt are renewable. Though, the main rainy season is Southwest Monsoon [June to September], southern Peninsular India also receives rainfall during the Northeast Monsoon [October to December] and cyclones/ depressions play vital role in getting widespread rains in pre-monsoon and post-monsoon seasons. All these follow cyclic variations as part of climate change. Thus, precipitation and snow melt are highly variable with space and time. Here climate change plays vital role in the year to year water availability over different parts of India. Therefore rains across the nation are not uniform and over years. India sees years of excess monsoons and floods, followed by deficit monsoons and droughts. This geographical and time variance in availability of natural water versus the year round demand for irrigation, drinking and industrial water creates a demand-supply gap that has been worsening with India's rising population – presently 1.21 billion may go up to 1.50 billion by 2050--, under new technological lifestyles – rapidly changing. In addition, with new technologies large part of the water is polluted through industries, domestic & agriculture. With the time this becomes major issue. Also, large part of fresh water is entering the Sea as we are unable to store and use them.

India receives 78% of the rainfall in four months [June to September]. This is estimated as 4,000 cubic kilometers [4 billion cubic meters (**bcm**)]. From this, 1.953 bcm goes as river flow in which 0.690 bcm is utilizable and 0.342 bcm goes to replenish groundwater. Recharge of groundwater from irrigation is around 0.086 bcm. That means, out of 0.432 bcm of water that goes to replenish ground water, around 0.396 bcm is utilizable. Of this 0.071 bcm goes for domestic, industrial & other uses and 0.325 bcm [90% of utilizable] goes to irrigation. Thus, the total Annual Utilizable Water Resources from surface and groundwater are around 0.690 + 0.396 = 1.086 bcm. However, water needs under domestic and industry are steadily increasing with passing of time. Global freshwater consumption quadrupled within the last 50 years; with the sector-wise withdrawals of water as 70% in Agriculture, 20% in industry 10% in domestic. However, sector-wise usage varied from developed and developing nations as well within them. **Table 2.1** presents Water Use by sector and country and **Table 2.2** present the expected water needs by 2050. From this it is clear that by 2050 the upper limit of water requirements exceed the water availability in terms of both the surface and the groundwater by 0.170 bcm [1.256 – 1.086]. This will increase further if we take the contamination part also.

Table 2.1: Sector-wise water use for few selected countries

Country	Sector-wise user in %		
	Domestic	Industrial	Agricultural
Canada	11	80	8
France	16	69	15
Germany	14	68	18
USA	12	46	42
Brazil	43	17	40
China	6	7	87
India	3	4	93
Source: FAO [1999] AQUASTAT estimates			

Table 2.2: Expected water needs by 2050

National Annual Water Requirement – projections in bcm:				
Year	Surface	Ground water	Total	Evaporation
(a) Industries				
2010	0.026	0.011	0.037	
2025	0.047	0.020	0.067	
2050	0.057	0.024	0.081	
(b) All activities				
2010	0.447-0.458	0.247-0.252	0.694-0.710	0.042
2025	0.497-0.545	0.287-0.298	0.784-0.843	0.050
2050	0.641-0.752	0.332-0.428	0.973-1.180	0.076
Replinishable groundwater – recharge in bcm as: Andhra Pradesh + Telangana States = 0.02003 + 0.01526 = 0.03529 and National as = 0.34243 + 0.089046 = 0.43189.				

2.3.1   Environmental Pollution Issues

Environment may be broadly understood to mean our surroundings and affects our ability to live on the Planet Earth—the air we breathe the water that covers most of the Planet Earth's Surface, the plants and animals around us, and much more. That is, the sum total of all surroundings of a living organism, including natural forces and other living things, that provides conditions for development and growth as well as of danger and damage. Environmental issues are harmful aspects of human activity on the biophysical environment. Environmentalism, a social and environmental movement that started in the 1960s, addresses environmental issues through advocacy, education and activism on current problems faced by the environment. Over population relates to carrying capacity. The global food production occupies 25% of all habitable land; responsible for 70% of fresh water consumption; 80% of deforestation; it is the largest single driver of biodiversity loss and land-use change that contributes to local & regional climate change. With the growing population, under globalization, the problems are increasing and of late they have taken a demonic size, which necessitated expert attention to be solved. As globalization paves its way across the world these problems no longer remain local problems but become international issues and there are numerous causes of these problems: created by man and can also be controlled by man. In recent years, scientists have been carefully examining the ways that people affect the environment. They have found that we are causing pollution, deforestation, and other problems that are dangerous both to the Planet Earth and to ourselves.

The term "pollutant" refers to any substance that, when introduced to an area, has a negative impact on the environment and its organisms. Pollutants can impact human health, air, water, land and entire ecosystems. Most sources of pollution result from human activity. There are numerous ways pollution is caused, such as industrial revolution, agricultural revolution and transport revolution. However, in urban areas waste associated with the population growth is another dimension of pollution. In the context of water pollution, transportation is not really an important issue.

The greater use of energy led to major problem of environmental pollution. With this energy entering the market several new polluting activities such as steel & iron, cement, bulk drugs manufacturing industries, etc., come up affecting environment. These added not only to air pollution components but also to water pollution components as well direct health impacts on life forms. All these are known point source pollution and thus with the progression of time invented controlling or reducing pollution technologies. As the manufacturing industries are profit driven, the control has not received its due place and thus pollution entering in to environment and became a major factor affecting nature in diverse ways.

In agriculture, newer technologies ensured greater production and thus technology was portrayed as a solution to all human problems, especially the problem of hunger and poverty. The use of chemical fertilizers under the so-called green revolution technology has risen astronomically not only in global prospective but also in India. In India the fertilizer use has increased from around 1 kg/ha to around 140 kg/ha in the last 60 years. Most of it is used under irrigated agriculture for crops such as paddy, wheat, sugarcane, cotton. According to World Bank data, per hectare fertilizer consumption in India, China, Bangladesh, Pakistan and Israel in 2007 stood at 142.3, 331.1, 171.2, 166.2 and 524.0 kg/ha, respectively. The chemical input technology created non-point source air and water pollution. Application of pesticides-insecticides created air pollution and the deposits of the same on ground contributed to water pollution also.

Water pollution is associated with agricultural runoff, which is surface water leaving farm fields because of excessive precipitation, irrigation, or snowmelt. Agricultural runoff is grouped into the category of nonpoint-source pollution because the potential pollutants originate over large areas and the point of entry into water bodies cannot be precisely identified. Nonpoint sources of pollution are particularly problematic because it is difficult to capture and treat the polluted water before it enters a stream. Because agricultural runoff is considered nonpoint-source pollution, efforts to minimize or eliminate pollutants focus on practices applied on or near farm fields. In other words, we usually seek to prevent the pollution rather than treating the polluted water.

2.3.2   River Water Pollution

River water pollution could be divided in to two parts, namely urban areas and rural areas. The pollution in urban water bodies are associated with unclean urban areas; no adequate water for dilution of polluted water that enters the rivers or water bodies; no sufficient Sewage Treatment Plants [STPs] to treat the wastewater generated to 100%; there are no provisions to stop wastewater joining the rainwater; and there are no strict regulations with religious activities.

Central Pollution Control Board [CPCB] of India published Water quality of rivers at interstate borders report has been put on its website on August 2015. CPCB has been monitoring the water quality of rivers at interstate borders since 2005 and is currently monitoring it at 86 locations across India. The current report covers 83 locations across 40 rivers. The report, based on water quality monitoring results from observations made between 2005 and 2013, revealed that 51 locations out of 83 are polluted in terms of biological oxygen demand (BOD); 57 locations have poor water quality on the basis of dissolved oxygen (DO); and at 70 locations, total coliform (TC) was more than the prescribed limit. BOD is a measure of the amount of dissolved oxygen required to completely oxidize available organic waste, while DO is the amount of oxygen dissolved in a given quantity of water. TC includes bacteria that are used as indicators to measure the degree of pollution.

In the case of rural areas, pollution from agricultural runoff is now considered the primary source. Pollutants in agricultural runoff include eroded soil particles (sediments), nutrients, pesticides, salts, viruses, bacteria, and organic matter. Through rivers the pollutants reach oceans and causing dead zones. The clear example to this is the mouth of Gulf of Mexico wherein the river Mississippi in USA enters the Ocean. That is, after green revolution technology agriculture became a major water polluting activity globally – both surface and ground as well both surface and ocean waters. Though it is a global problem nobody pays attention to this.

Agriculture is the major source of non-point pollution and this must be tackled through changes in technology. This is discussed by **Reddy [2011]**. UN World Economic Social Survey chapter on sustainable agriculture under small farm holders, FAO in 2011 released an edited book titled "Climate Change and Food System

Resilience in Sub-Saharan Africa" and several other international reports suggest that ecological agriculture holds significant promise for increasing the productivity in developing countries. The use of chemical fertilizers is steadily decreasing under this system. The green revolution system has shown that increase in yields doesn't necessarily translate into food security. That is, technological strategy does not guarantee food security or even social security. The so-called success of the green revolution system was due to heavy government incentives in terms of providing subsidies, building infrastructures and providing guarantee for credits. It is not a sustainable agriculture. In India, the Ministry of Agriculture is promoting organic farming in the country under National Project on Organic Farming, National Horticulture Mission, etc. NGOs in collaboration with farmers are working over different parts of the country. Farmers achieved remarkable yields. This needs more thrust. It is not only going to help governments but also to the nature as well people in general in terms of health and farmers in specific in terms of improving household economy and nutrient security. At the same time it acts environment friendly. The importance of organic farming has been recognized globally but moving at snails speed. To compensate in production if any could be accomplished through bringing rainfed areas under irrigation through dams and inter-linking of rivers. The fertilizer barons' are having hectic activity at UN so that pollution associated with fertilizers does not enter in to COP21 Paris December meet.

Musi River pollution: A case study: Musi River is one of the tributaries of Krishna River, originate in Vikarabad's Ananthagiri Hills via Hyderabad the capital city of Telangana State, after travelling around 250 km meets the Krishna River at Vazeerabad [Wadapally]. Originally river Musi was used for drinking water purpose en route and also for irrigating around 30,000 ha of land. There are about 30 Anicuts [known as Katwas] on river Musi. Now it is a cesspool of poison. The major cause is government's apathy in protecting the water of the river and other water bodies in the catchment area. Governments thought it is easy to get water from faraway places by spending thousands of crores through lift irrigation rather than protecting the water resources available through simple gravity and finally Turing it in to polluted water. The twin cities of Hyderabad and Secunderabad used to discharge around 275 MLD of domestic sewage into the Musi River by around 1960's. To solve the problem of this pollution a sewage treatment plant was established at Amberpeta. Of the 275 MLD, around 160 MLD is treated partially [only primary treatment] and discharged in to river Musi along with the remaining 115 MLD of untreated sewage.

Industrial areas have been developed in Musi River basin after 1960. Major percentage of industries in these industrial areas consists of Synthetic Organic Chemicals [bulk drugs and intermediates], Oil refineries, Textiles, Tanneries, Electroplating units, and Distilleries. There are about 182 industries discharging their treated and partially treated and untreated effluents directly and indirectly to river Musi. Also the solid wastes generated by these industries dumped everywhere in the Musi River Basin joins Musi River with rains. With the passing of time, three Central Effluent Treatment Plants [CETPs] – PETL, JETL & IDPL – were established to treat the industrial effluents. These are treating partially as the technology was not available to them – mainly diluting industrial effluents with domestic waste water/sewage. PETL used to discharge the partially treated effluents in to Nakkavagu is now discharging in to Musi River through an 18 km pipeline that joins with JETL & IDPL partially treated effluents. In addition to these partially treated effluents, untreated effluents coming from excess production and the industries that have no treatment facilities are being dumped in to water bodies in the respective industrial areas and the solids in to forest areas – low lying areas and finally joining Musi River through rainwater runoff. All these have contaminated groundwater.

Simultaneously the riverbed lands of river Musi were also encroached by around 50%. With the growth of the population around 2000 MLD of sewage is generated. Sewage enters the Musi River from 18 sewer lines from different parts of the city. However, large part of it is dumped in to water bodies and as a result the entire groundwater in those zones has become contaminated water. All the sewage is finally entering the Musi River along with the industrial effluents. Only around one-third of it is treated at STPs – Hussainsagar – 20 mld; Amberpet – 339 mld; Nagole – 172 mld; Nallacheruvu – 30 mld; Zia Guda/ Nandimurisalai – 51 mld [total of 592 mld] plus few more STP at selected water bodies were established. However, large parts of them are either non-operative or working now and then. Also, sewage is discharged directly in to rainwater drains and that joins

Musi directly through rainwater. Now Musi River is a cesspool of poison. Using this water around 2.5 lakh acres of land is cultivated. The food produced from such water is supplied to twin cities. Now, the government has released the contaminated water from Hussainsagar directly in to Musi River to establish real estate ventures around Hussainsagar Lake. In Andhra Pradesh & Telangana states, river beatification means establishment of real estate ventures under the disguise of eco-tourism an "adda" for anti-social activities.

Osmansagar and Himayatsagar Reservoirs: The Mohd Quli Qutub Shaw founded Hyderabad on the banks of Musi River in 1591. On the advice of Sir M. Visvesvaraya the legendry Engineer of India, Himayatsagar and Osmansagar reservoirs were built on Musi and its tributary Easi to contain floods & to provide drinking water to twin cities after severe floods in September 1908. Osmansagar was built in the year 1913 and Himayatsagar was built in 1927. These two reservoirs have a supply capacity of around 85 MCM, which was just sufficient for population of 1.25 million in 1961. With them, the free flow of rainwater in to Musi reduced drastically.

On 19th June 2015 media carried out a report relating Himayatsagar & Osmansagar Lakes that provide drinking water to Capital city. From these reports it is clear that the government is planning to build four sewage treatment plants [STPs] to treat the effluents entering in to these two lakes and later dump the treated water in to these two **drinking water** lakes. Nowhere in the world is such activities carried out.

In 1989, GO50 was issued by the government to protect the quantity of water flowing in to these two lakes. In 1994 another GO192 was issued by the government to protect the quality of water flowing in to these two lakes. In 1996 these two GOs were merged and brought out a single GO111 of MAUD/GoAP to protect the quantity & quality of water flowing in to these two lakes. The GO111 prohibits various developments within the 10 km radius from the two lakes full tank levels [FTL], which are the main sources of drinking water supply to Capital city by gravity. It states that "prohibits polluting industries, major hotels, residential colonies or other establishments that generate pollution, in the catchment of the lakes up to 10 km from FTL"; "residential developments in residential use zone may be permitted with 60%

of the total area kept as open spaces and roads in all layouts in the villages of prohibited catchment area [identified 84 villages]"; restricts the FSI to 1:0.5 in the catchment area. The measures will ensure that the 90% of the area remains under agriculture, as per the prevailing practice and ensure protection of the lakes"; and "the land use of about 90% of the catchment area is classified as recreational and conservation use in the Master Plan. The Hyderabad Urban Development Authority should take action for classification of this 90% of the area as agriculture, which is inclusive of horticulture and floriculture". The GO also puts restrictions on air polluting industries on downstream of the lakes and included GO 50 dated 28-1-1989 relating to quantity of water. This GO lists the departments that need to protect the lakes as per GO111.

Government in violation of this GO, within few days of its issue, granted permission to a pollution potential industry. Environmental groups approached the Hon'ble Supreme Court. The Hon'ble Supreme Court in its order dated 1st December 2000, ruled that, "It is, in our view, not humanly possible for any department to keep track whether the pollutants are not spilled over. This is exactly where the **precautionary principle** comes in to play. On the basis of scientific material obtained by the court, we hold that the pollution control board could not be directed to suggest safeguards and there is every likelihood that safeguards could fail either due to accident or due to human error". Yet the present government in violation of this order wanted to build STPs and treated effluent will be dumped in to two reservoirs of drinking water. This shows the scant respect our politicians/bureaucrats have on court directions leaved alone their own orders, even in the case of drinking water reservoirs.

With this judgment all the polluting industries were closed. But, unfortunately, the government itself started time and again violating "GO111 and Supreme Court judgment", allowed to come up certain activities [Shamshabad Airport, ORR-Phase-I, etc] and majority by private groups built activities. All these have severe impact on quantity and quality of water flowing in to these two lakes. I on behalf of environmental groups filed a PIL in 2007 [W.P.No.9386/2007]. On this the government submitted the list of violating activities and issued a Memo No.14046/I1/07 dated 12-10-2007. In this memo it is observed that "In the reference 3rd cited [From the Advocate General. A.P., Letter No. 423/07 dated 11-10-2007], the learned Advocate General of Andhra Pradesh has informed that the Writ Petition has come up for hearing before the Division Bench of Hon'ble High Court of Andhra Pradesh on 11-10-2007 and the Division Bench took serious view about the constructions which are coming up in the above mentioned area and directed the Government to issue appropriate orders to all concerned departments, stalling the entire construction activity within the upstream area of Himayatsagar and Osmansagar Lakes. Accordingly, Government hereby instruct all the concerned authorities – which are associated with regulating the construction and development activity within the catchment area of Himayatsagar and Osmansagar Lakes not to allow any construction, layout and non-agriculture development activity within the upstream area of Himayatsagar & Osmansagar Lakes in tune with G.O.MS.No.111M.A. Dated 8-3-1996". Government also brought out a G.O.MS.No.157 dated 6-4-2010 constituting Lake Protection Committee. Here it is pertinent to note that "Government have examined the proposal and keeping in view the reports of the Metropolitan Commissioner, HMDA and in compliance of the orders of Hon'ble High Court in W.P.No.9386 /2007, dated 2-2-2010, have decided to constitute a Lake Protection Committee for preservation and protection of Lakes in HMDA area pending consideration of enactment of separate legislation for constitution of Lake Protection Authority". Authority has legal power but committee has no such powers.

With all these, while the PIL is still pending with the Hon'ble High Court, the violations have been doubled-tripled. The water flows reduced by more than 50% and the concentrations of pollutants tripled. At this time the vested groups approached the Hon'ble High Court in the past they tried their best to kill the GO111 for around one thousand crores – on pollution issue and got the order to pollution control board from the court to construction of STPs. Here neither the Court's Registrar's office nor the pollution control board has brought to the notice of the Respected Bench on the Supreme Court's order of precautionary principle or the W.P.No.9386/2007 that is pending in the court. It appears that this was carried out secretly at the behest of vested groups for a price. Now, the government is trying to implement the same.

In this connection, presenting above details, I sent a letter to the Chief Secretary, asking him "In the capital city around 2000 million litres per day [mld] of wastewater is generated daily. Is the government is willing to use this water after passing through STPs in government offices including Secretariat and government residential quarters including Judges Colonies???" However, as usual there was no response. While this is the case, The Times of India [28th August 2015] report states that "The Cyberabad police are donning a new role. The lathi-wielding men in khaki are drawing up plans to protect century old drinking water sources Osmansagar and Himayatsagar from encroachments. Cyberabad police commissioner has floated a proposal 'Osmansagar and Himayatsagar Lakes Protection & *Development project*' to protect both the water bodies from encroachment". On this I sent a letter to the Cyberabad Police Commissioner and also met him personal requesting him not to interfere with the GO111 and Supreme Court order. What they wanted to do is to kill the GO111 and create a ring road around it that help anti-social activities and the inturn help the illegal activities within the FTL to 10 km radius in the catchment area of the two lakes. This shows the scant respect for water resources. With all these violations, Mahanadi a Siva temple near Nandyala in Kurnool district of AP presents a classical case of fresh water flows for centuries.

2.4   Irrigation Systems in India

2.4.1   General

Irrigation is the artificial exploitation and distribution of water at project level aiming at application of water at field level to agricultural crops in dry areas or in periods of scarce rainfall to assure or improve crop production. All the different sources of irrigation in India are divided into two major divisions; viz. Flow irrigation and Lift irrigation. The water of a reservoir or tank usually remains at a higher level, and when a channel is connected to it, water automatically flows down the channel which serves the purpose of a canal for irrigating the land. In this case the water level remains higher than the fields. Such irrigation is known as the flow irrigation and it is generally possible in the plain areas.

With the farm lands lie at a higher level and the canals or tanks lie at a lower level, it becomes necessary to lift the water by pump etc., to irrigate land. Water is lifted from wells and tanks by a crude country method and from tube-wells by pumps for irrigation.

Nowadays the ground water is used for irrigation by lifting it by means of electric or diesel pump sets. Water is also lifted from wells tanks or rivers by pumps and irrigation is done through channels. This method of irrigation is known as the Lift Irrigation. Nowadays sprinkle irrigation is being very much popular as more land can be irrigated with less water in this method.

The Irrigation Projects of India are classified into three types according to their capacity of irrigation. They are (i) Major Irrigation Projects, (ii) Medium Irrigation Projects and (iii) Minor Irrigation Projects. Irrigations in India are carried out in three different ways according to their sources, such as (i) by canals, (ii) by wells, and (iii) by tanks. Out of the total area under irrigation, 40 percent are irrigated by canals, 40 per cent by wells and 12 percent by tanks. The rest 8 per cent of land are irrigated by other methods.

Irrigation by Canals: This is the most convenient method of irrigation. About half of the total area under irrigation by canals is situated in Punjab, Haryana, Uttar Pradesh and Andhra Pradesh. It is easy to dig canals in these areas since the land is level and soil soft. There are two types of canals; such as: perennial canals and inundation canals. Artificial reservoirs are created by constructing anicuts, barrages or dams across rivers for perennial canals. Irrigation is being done in the Mahanadi delta area by constructing barrages at Naraj and at Jobra of Cuttack city across the Mahanadi and at Choudwar across the river Birupu. When there is excessive flow of water in the rivers in flood, the extra water flows in the canals rising from those rivers. Such canals are effective only during floods; hence those are known as the inundation canals. This type of canal is very few in number in the country, more in Punjab than elsewhere.

There are many perennial canals in different regions of the country and the most famous of those are the Upper Bari Doab Canal and the Sir hind Canal in Punjab, the West Yamuna Canal and the Chakra Canal in Haryana. The Chakra Canal is the largest canal of the country. This canal serves the purpose of irrigation in the states of Punjab and Haryana. The Rajasthan Canal (The Indira Gandhi Canal) of Rajasthan is the longest canal of Asia. The northwestern part of Rajasthan is being irrigated by it. The other important canals are the Shard Canal, the Beta Canal, the Upper Ganga and the Lower Ganga Canals of Uttar Pradesh. Many canals have been dug out of the rivers Krishna, Godavari und

Tungabhadra of Andhra Pradesh. The other important canals are the Son Canal of Bihar, the Damodar Canal of West Bengal, the Mahanadi and the Rushikulya Canals of Orissa, the Mettur and the Periyar Canals of Tamilnadu; the Krishnarajsagar, the Tungabhadra and the Ghataprava Canals of Karnataka.

Irrigation by Wells: The rain-water sinks down easily in the areas where the soil is soft and porous. So water is available at a lower depth when wells are dug and it helps irrigation. Primarily irrigation is carried on by wells in the western part of Uttar Pradesh, some parts of Bihar and in the blank cotton soil area of the Deccan. In addition to it, in the coastal strip of Tamilnadu and Andhra Pradesh, some parts of Rajasthan, Haryana and Gujarat irrigation is also carried on by wells. Some or the other type of lift irrigation is required-for using the well-water for irrigation. Old methods like inot or ieiida are still practiced in many areas. Power-driven pumps have become very popular in most parts. In some areas cattle or camels are used to lift water by the Persian wheels. The power-driven (electric or diesel) pumps can also lift water from a much greater depth from tube-wells. Nowadays wind mills also lift water from the wells for irrigation purpose. Irrigation by wells is more expensive, so more profitable farming of vegetables is carried on in those areas.

Irrigation by Tanks: Tank irrigation is the most feasible and widely practiced method of irrigation all over the Peninsula, where most of the tanks are small in size and built by individuals or groups of farmers by raising bonds across seasonal streams. The soil of this plateau is hard and stony and its land is undulated and so, it is not easy to dig canals or wells in those areas. There are big tanks which have been created by raising high bonds on one side of the valley of hills. Small channels are dug out of both the sides of the tanks to irrigate lands. There are big reservoirs like Nizam Sagar, Osman Sagar, Hussain Sagar, Krishna raj Sagar etc., in the peninsular India. Besides, in rural areas of the Peninsula there are large numbers of small tanks for irrigation, but such tanks dry up during acute drought period and don't help in irrigation.

Underground Water Resources: A huge quantity of water has been stored under the ground since long. Now an organization named the Central Underground Water Board has been set up in order to utilize this water. A map has been prepared by this organization after surveying all over the country. This map shows the regions where underground water is easily available and at what depth. Nowadays pure drinking water is being supplied by deep bored tube wells in the rural areas where pure drinking water is not available and in the coastal strip of Orissa having drained water. Lift irrigation is being carried on by sinking deep bored tube wells in the areas having scarcity of water for agriculture. People had not such idea regarding the underground water before India attained independence. So the underground water resources had not been developed.

Multipurpose River-valley Projects: Many multipurpose river valley projects have been developed in our country in order to utilize the vast water-resources of our rivers. Many purposes can be solved by creating reservoirs by constructing strong dams and embankments or bonds in the river beds. Flood control, irrigation, generation of hydro-electricity, navigation, soil conservation, afforestation, pisciculture, water supply etc., have been the chief aims of these multipurpose projects. So these projects have been turned as Multipurpose River-Valley Projects. In addition to this, these projects have also been spots of tourist interest to attract tourists. Owing to generation of hydro-electricity, power supply has been cheap and convenient and growth of industry has been possible in our country.

Water Management: The most important physical elements of an *irrigation project* are *land* and *water*. In accordance with the propriety relations of these elements there may be different types of water management, such as the communal type, the enterprise type, the utility type, etc. Until the end of the 19th century the development of irrigation projects occurred at a mild pace, reaching a total area of some 50 million ha worldwide, which are about 1/5 of the present area? The land was often private property or assigned by the village authorities to male or female farmers, but the water resources were in the hands of clans or communities who managed the water resources *cooperatively*. The enterprise type of water management occurred under large landowners or agricultural corporations, but also in centrally controlled societies. Both the land and water resources are in one hand.

Large plantations were found in colonized countries in Asia, Africa, and Latin America, but also in countries employing slave labor. It concerned mostly the large scale cultivation of commercial crops such as banana, sugarcane and cotton. As a result of land reforms, in many countries the estates were reformed into cooperatives in which the previous employers became members and exercised a cooperative form of land and water management. The utility type of water management occurs in areas where the land is owned by many, but the exploitation and distribution of the water resources are managed by (government) organizations. After 1900 governments assumed more influence over irrigation because:

- water was increasingly considered government property owing to the increasing demand for good quality water and the reducing availability;
- governments embarked on large scale irrigation projects as they were considered more efficient;
- the development of new irrigation schemes became technically, financially and organizationally so complicated that they fell outside the capabilities of the smaller communities;
- the import and export policies of governments required the cultivation of commercial cash crops whilst, by controlling the water management, the farmers could be more easily guided to plant these kind of crops.

The water management signified a large subsidy on irrigation schemes. From 1980 the operation and maintenance of many irrigation projects was gradually handed over to water user associations (WUAs) who were to assume these tasks and a large part of the costs, whereby the water rights of the members had to be respected. The exploitation of water resources via large storage dams - that often provided electric power as well - and diversion weirs normally remained the responsibility of the government, mainly because environmental protection and safety issues were at stake. In the past, the utility type of water management witnessed more conflicts and disturbances then the other types.

2.4.2   Historical Perspective

Civilizations have been dependent on development of irrigated agriculture to provide agrarian basis of a society and to enhance the security of people. Civilizations have risen and fallen with the growth and decline of their irrigation systems, while others have maintained sustainable irrigation for thousands of years. Archaeological studies have identified evidence of irrigation in Mesopotamia and Egypt as far back as the 6th millennium BC; and in the 'Zana' Valley of the Andes Mountains in Peru from around 3-4th millennium BC. The Indus Valley Civilization in Pakistan and North India (from 2600 BC) also had an early canal irrigation system. Sophisticated irrigation and storage systems were developed, including the reservoirs built at Girnar in 3000 BC. There is evidence of the ancient Egyptian Pharaoh Amenemhet-III in the twelfth dynasty (about 1800 BC) using the natural lake of the Fayum as a reservoir to store, as the lake swelled annually surpluses of water for use during the dry seasons as caused by the annual flooding of the Nile. The Qantas, developed in ancient Persia in about 800 BC is still in operation. The irrigation works of ancient Sri Lanka, the earliest dating from about 300 BC, in the reign of King Pandukabhaya and under continuous development for the next thousand years, were one of the most complex irrigation systems of the ancient world. In fifteenth century Korea of the World's first water gauge (woo ryang gyae) was discovered in 1441. It was installed in irrigation tanks as part of a nationwide system to measure and collect rainfall for agricultural applications. With this instrument, planners and farmers could make better use of the information gathered in the survey.

Ministry of Water Resources Government of India, on its website briefly explains the history of irrigation development in India which can be traced back to prehistoric times. Vedas, ancient Indian writers and ancient Indian Scriptures have made references to wells, canals, tanks and dams. These irrigation technologies were in the form of small and minor works, which could be operated by small households to irrigate small patches of land. In the south, perennial irrigation may have begun with the construction of the Grand Anicuts by the Cholas as early as second century to provide irrigation from the Cauvery River.

The entire landscape in the central and southern India is studded with numerous irrigation tanks which have been traced back to many centuries before the beginning of the Christian era. In northern India also there are number of small canals in the upper valleys of rivers which are very old. We can find excellent systems built by Rishis in Andhra Pradesh along the periphery of Nallamala forests under Shiva temples like Srisailam and Mahanadi. Giyasuddin Tughluq (1120-1250) is credited to be the first ruler who encouraged digging canals. Fruz Tghluq (1351-86) is considered to be the greatest canal builder. Irrigation is said to be one of the major reasons for the growth and expansion of the Vijayanagar Empire in southern India in the fifteenth century. In Andhra Pradesh we can see Cumbum Cheruvu still intact – the second largest in Asia (It is said that Rishi Jamadagni built it and later restored it by Oriya King).

2.4.3   Source-wise Irrigation

Irrigation has acquired increasing importance in agriculture the world over. From just 8.0 million hectares (Mha) in 1800, irrigated area across the world increased fivefold to 40 Mha (13.4 Mha in India) in 1900, to 100 Mha in 1950 and to just over 255 Mha in 1995. With almost one fifth of that area, India has the highest irrigated land in the world today. At the global scale 278.8 Mha (689 Million acres) of agricultural land was equipped with irrigation infrastructure around the year 2000. About 68% of the area equipped for irrigation is located in Asia, 17% in America, 9% in Europe, 5% in Africa and 1% in Oceania.

Close to nineteenth century according to sources of irrigation: Canals irrigated 45%, wells 35%, tanks 15% and other sources 5%. During 1910 to 1950 the growth rate of irrigation was estimated at 2.0% per annum for government canal irrigation, 0.54% per annum for well irrigation and 0.98% per annum in respect of irrigation from all sources. At the time of Independence net irrigated area in India under British rule which include Bangladesh and Pakistan was 28.2 Mha. After patrician net irrigated area in India and Pakistan being 19.4 Mha and 8.8 Mha respectively. **Table 2.3** presents the % of the total area for India and AP under different sources of irrigation during 1960-61 and 1999-2000.

India's irrigation development in this century, and particularly after independence, has seen large number of large storage based systems, all by the government effort and money. Post independence has seen more than 60% of irrigation budgets going for the major and medium projects. India's ultimate irrigation potential was estimated at 139.9 Mha, comprising of 58.46 Mha through major and medium irrigation schemes and 81.43 Mha from minor irrigation schemes. Recently some positive steps were also taken to long-awaited inter-basin water transfer, aiming at adding 35 Mha to India's irrigated area.

Table 2.3: Share of different sources of irrigation in India and AP

Source	India [%/Mha]		Andhra Pradesh [AP] [%/Lha]	
	1960-61	1999-00	1960-61	1999-00
Canals	42.05/10.37	31.29/17.45	45.75/13.31	37.27/16.34
Tanks	18.50/04.56	05.18/02.94	39.57/11.51	14.85/06.51
Wells	29.56/07.29	57.81/24.70	11.31/03.28	43.34/19.00
Others	09.89/02.44	05.73/02.93	03.40/00.99	04.54/01.99

2.4.4 Groundwater

India uses around 25% of the world's groundwater. Out of the total 5723 groundwater blocks in India, 30% are already in the danger zone due to overexploitation. This may go up to 60% by 2025. In India, 33% of fresh water use comes from groundwater. The over exploitation of groundwater not only reduced the soil quality but also reduced drastically the area cultivated under pumps – In AP in 70s it was 2.5 ha per pump and now it is around 0.5 ha. From **Table 2.3** it is clearly evident that the sources of recharging groundwater, namely tanks have drastically come down.

The annual replenishable groundwater resources for the entire country are 0.433 bcm. The overall contribution of rainfall to the country's annual replenishable groundwater resource is 67% and the share of other sources, including canal seepage, return flow from irrigation, seepage from water bodies and water conservation structures taken together is 33%. However, in AP., Haryana, Delhi, Punjab, J&K, Jharkhand, TN, UP, Uttarkhand and UT of Pandichery this is more than 33% mainly because of canal seepage and intensive irrigation.

The southwest monsoon being most prevalent contributor of rainfall in the country, about 73% of country's annual replenishable groundwater recharge takes place during kharif period of cultivation by keeping 0.034 bcm as allocation for natural discharge during non-monsoon season. However, availability of groundwater is widely variable across the length and breadth of the country. Groundwater resources availability, utilization and categorization are given in **Table 2.4**. The annual groundwater available for utilization for the country is 0.399 bcm. However, the total annual draft is 0.231 bcm, out of which 0.213 bcm is used for irrigation and 0.018 bcm issued for domestic and industrial use. In general, the irrigation sector remains the main consumer of groundwater (92%).

Table 2.4: Groundwater resources availability, utilization and categorization

Parameter	AP	India
Annual replenishable Groundwater source [bcm]	36.50	433.02
Natural discharge during Non-monsoon season [bcm]	03.55	033.77
Net groundwater availability [bcm]	32.95	399.25
Annual groundwater draft [bcm]	14.90	230.62
Stage of groundwater development (%)	45	58
Overexploited [blacks]	219	839
Critical [blacks]	77	226

2.5 Local Political Syndrome: A Case Study

General: Andhra Pradesh State was bifurcated in to Andhra Pradesh and Telangana States. Before bifurcation of AP state, Projects were built to utilize Krishna River water as per the tribunal award [Justice Bachawat Tribunal]. Among these the important ones are Jurala, Srisailam & Nagarjunasagar Dams. All the three provide water for irrigation and drinking/industry and as well for hydro-power production. However, in the case of Godavari River water use, projects construction hasn't preceded as required though tribunal allocated water is available. This is basically because the river is at a lower level than the agriculture area by several hundreds of meters and thus it has to be lifted using power. Prior to 1990 power production was not that high to use it for lift irrigation. Where ever water is available through gravity some projects were completed and are in use. Thus, major share is unutilized and as a result cost of building projects gone up and up with the time. Dr. Y. S. Rajashehara Reddy [hereafter referred as Dr. YSR] took over as Chief Minister of AP in 2004, and he proposed to build around 80 projects under Jalayagnam. This includes the projects to utilize the surplus water/flood water under Krishna River as per Justice Bachawat Tribunal. When Dr. YSR initiated these projects, opposition parties went on ranting "Jalayagnam is Danayagnam". If we look at projects initiated under Jalaygnam in Telangana region [new state]:

A report in Andhrajyothi June 2014 state that in Telangana under Jalayagnam 25 projects were initiated to irrigate 22,77,416 acres at a cost of about Rs. 21,506 crores. So far around 14,321 crores were spent on them. After Dr. YSR death in 2009, the subsequent governments failed to speeding up the construction activity as, if they are completed the credit goes to Dr. YSR – it is a political game at the cost of development. The report stated that to complete the projects Rs. 11, 440 crores is needed and in which Rs. 7,185 crores comes as a part of original estimate plus an addition of Rs. 4,255 crores [escalation costs]. However, the new government is playing politics and wanted to change the projects. These projects include under Krishna River as well Godavari River. One of the major projects initiated on Godavari River is Pranahita-Chevella project that is expected to meet the drinking and irrigation needs in 7 districts out of ten districts of Telangana and thus bringing down drastically the cost of water grid project initiated at Rs. 40,000 crores. Telangana government go on announcing projects running in to lakhs of crores without proper assessment in an integrated manner that will burden the common man with the passing of time with loans taken left and right. Through integration of all water related projects the cost could be drastically brought down but this is not their objective. For them projects means percentages.

Another example on this can be seen from Andhra Jyothi dated 22nd September 2015 in which published an article relating Thotapalli Reservoir. The report states that P. V. Narasimha Rao, Prime Minister of India in 1993 laid a foundation stone for Thotapalli Reservoir linking with flood flow channel of Sriramsagar. The flood flow channel of around 60 km was completed. Dr. YSR in 2006-07 has taken up the works Thotapalli, Gauravelli, and Gundavelli reservoirs by linking to the flood canal. This group of reservoirs was linked to Pranahita-Chevella lift irrigation project and as a part 12 km tunnel was completed. Guauravelli and

Gundavelli reservoirs work of 90% was completed.   Due to builders negligence and non-procurement of land for Thotapalli reservoir was not completed. But now the present government cancelled this project and trying to do real-estate & industrial-estate business.

Lift Irrigation projects under Krishna River: Bachawat Tribunal allocated 811 TMC of water to undivided Andhra Pradesh in which 11 TMC is part of return flows and 5 TMC to Chennai drinking Water.   The 800 TMC will be available only in 75 years in 100 years. The remaining 25 years the water available will be less than 800 TMC and accordingly the return flows will be less than 11 TMC and yet 5 TMC has to be given for Chennai drinking water. In 800 TMC on an average locally available part is 350-400 TMC [major part is associated with cyclonic activity] and the rest has to come from Upper Riparian States, namely Karnataka and Maharashtra. From this, the water allocations to Telangana is 298.96 TMC in which **20 TMC for Bhima lift irrigation** will be available only after Krishna Delta modernization [16th April 1996 CWC approved]. The modernization has not been done as allocations for the execution were meager. So, at present available water is 278.96 TMC.   The project-wise allocations in TMC are: The major share was allocated to Nagarjunasagar dam – 105.70; and next in order is small irrigation sources – 89.15. The rest are: Jurala – 17.84; RDS – 15.0; Musi – 9.40; Hyderabad drinking water – 5.70; Chennai drinking water – 1.67; Srisailam evaporation – 11; Okasetti vagu – 1.90; Kotipalli vagu – 2.00; Lankasagar – 1.00; **Koyalsagar – 3.90; Dindi – 3.50**; Paleru – 4.00; Vyra – 3.70; & Pakala Lake – 2.60.

The Bachawat tribunal in lieu of 25% of the deficit years allowed undivided AP to utilize the flood waters over 811 TMC when they are available for drier areas.   To utilize the flood water when available undivided AP government proposed 8 projects to meet the water needs in drier areas – the present Telangana CM was part of TDP regime at that time. Though they were proposed several decades back money was not allocated to execute them except laying foundation stones by TDP Chief Minister N. T. Rama Rao [NTR] and second time by Chief Minister Chandrababu Naiudu [BN] of TDP to favour illegal use of surplus water beyond Srisailam Dam.   After Dr. YSR took over as CM in 2004 the funds were allocated to these projects under Jalayagnam. After his death, the successive governments failed to complete them by allocating sufficient funds. According to a press release by the irrigation minister [Sudarshan Reddy from Nizambad] in 2013, the water allocations for these 8 projects are (in TMC) 227.5 and of which 77 for the three Telangana projects, namely S.L.B.C – 30, Kalvakurthi – 25 & Nettempadu – 22. We must not forget the fact that this water is available only once in two years or less under Bachawat tribunal. With the Brijesh Kumar Tribunal order this will be once in four years or may not even once in hundred years – this is discussed in **Chapter 3**.

In the same press release by the irrigation minister hasn't mentioned Palamuru lift irrigation project with 70 TMC [now some reports say 90 TMC taken from Srisailam dam through lift irrigation and of which 20 for drinking and 70 for irrigation], but the present government announced this along with 30 TMC each for Dindi & Koyalsagar lift irrigation projects though they were allotted 3.5 & 3.9 TMC from allocated water. Then where from this 150 TMC of water comes is a big question. And yet the government initiated activities on Palamuru project. This shows how politicians play game with the lives of people – vote bank politics. I brought to the notice of Chief Secretary of Telangana State on this.

A "Front page" report by Eenadu [27-9-2015] stated that Telangana government is going to utilize 197 TMC from surplus water. That means 120 TMC over and above 77 TMC. If we look at water allocations to Telangana [298.96 TMC] from 811 TMC is around 36.9%. By following the same analogy the total surplus water planned to use by AP + TS comes as around 533.9 TMC [= 197/0.369].

At 2060 TMC water availability in the river, in 25% of the years AP will get less than 811 TMC and thus no surplus water. To compensate this deficit 150 TMC of water is allowed to store during surplus years to meet the deficit to projects that were distributed 811 TMC. That means the project built for surplus water use, will not get water if the river flows are less than 2210 [2060 + 150] TMC. The 2210 TMC is available at 65% or 65% probability levels under 78 [1894-95 to 1971-72] or 114 [1894-95 to 2007-08] years data series. That means, in 35% or 35% of years the projects built on surplus water will not get water.

Projects built on surplus water to utilize 227.5 TMC of water, river flows must have 2437.5 [= 2210 + 227.5] TMC. This value on the above mentioned two probability curves meet at 42.5% or

47.5% probability levels. That means, 227.5 TMC of water will be available in 42.5% or 47.5% of the years only. That means, once in two years this water may be available. Between 65%/65% and 42.5%/47.5% of the years only partial part of 227.5 TMC may be available. In the past few decades in the absence of projects to utilize surplus water, after Srisailam dam, illegal use comes around on an average this quantity only.

Projects built on surplus water to utilize 553.9 TMC of water, river flows must have 2743.9 [= 2210 + 553.9] TMC. This value on the above mentioned two probability curves meet at 25% or 27.5% probability levels. That means, 553.9 TMC of water will be available in 25% or 27.5% of the years only. That means, once in three/four years this water may be available. Between 65%/65% and 25%/27.5% of the years only partial part of 553.9 TMC may be available.

Here one important point that must be kept in mind is that all the water that is going in to the Bay of Bengal cannot be considered as surplus water and also more than 50% of water allocations must come from upper riparian states, namely Karnataka and Maharashtra. With this the probability levels presented above with 227.5 & 553.9 TMC of surplus water use will be further reduced due to certain practical issues/on several accounts. For example (1) if the upper riparian states use illegally excess water [already several illegal projects were built]; (2) with floods associated with cyclonic activity between Nagarjunasagar and Prakasham barrage that count for surplus going in to the Bay of Bengal as the Prakasham barrage capacity is low and thus to avoid any damage to barrage, water will be released in to Bay of Bengal after crossing the fixed level; and (3) with the climate cycle – during the below/above the average rainfall cycle the probabilities go down/up. This scenario under Justice Brijesh Kumar Tribunal award is quite different. They may be once in ten years or zero years – top TS Think Tank supported this award and because of this, to weaken the SC case on this award TS filed case on sharing of water between AP & TS [which I brought to your notice through my mail dated 29th August 2015].

Pranahita-Chevella Project: Though in Godavari River, Bachawat Tribunal allocated 1480 TMC and equal amount available under flood water. In this river around 3000 TMC is entering the Sea on an average per year due to non utilization of water by undivided

AP and upper riparian states. Dr. YSR proposed projects to utilize the water allocated by Bachawat Tribunal. He proposed bring 30 TMC of Godavari water to Hyderabad to serve the drinking water needs – this is part of allocated water only and not flood water like Palamuru-Ranga Reddy 20 TMC for Hyderabad drinking water – and through Pranahita-Chevella project with 160 TMC to serve drinking and irrigation needs in 7 districts of Telangana. Polavaram is in Andhra Pradesh State – this is discussed in **Chapter 4**.

Now the Telangana Government after one and half year in power wants change all those projects initiated by Dr. YSR at huge cost. Already thousands of crores were spent on all these projects and with some money spending the projects excluding Pranahita-Chevella are ready to use for drinking and irrigation. This will reduce the needs of Water Grid to a maximum extent. A request for national status to Pranahita-Chevella is pending with the Central Government – Telangana Government pursued this matter with the Central Government. It appears that the government is worried on four counts, namely (1) if these projects are completed as planned, Dr. YSR get the credit, (2) The percentages on contracts will not come, (3) If South Telangana Politicians raise the issue of separate state then they may have to face the similar to Krishna problem, and (4) If these projects are completed in time as planned the people may not vote to ruling party and the money required to water grid project may come down to less than 10,000 crores instead of 40,000 crores. The government must present a white paper on these.

Kakatiya Mission: On Kakatiya Mission – myths & realities, the author presented an article in the daily newspaper "Andhra Jyothi, 23/6/2015". After AP was formed, the governments built dams more particularly on Krishna River and provided water for drinking – irrigation and industry. People are enjoying the fruits of this achievement. Before the state formation, rulers of the day built tanks that helped drinking – irrigation needs of the people. The tanks helped recharging the groundwater. Traditionally farmers with the help of local-village level revenue network used to look after the tanks in terms of de-silting and repairs to the tank bund – up keeping of the tanks. Later this system became defunct with the destruction of village level revenue network by TDP government headed by NTR in which regime present Telangana CM was part.

According to Telangana government there are 46,477 tanks in Telangana and of which 22,950 are in Krishna River Basin and 23,497 in Godavari River Basin and allocated Rs. 25,000 crores to restore these tanks. According to government estimates, around 255 TMC of water is available from these tanks and using these water 20 lakh acres could be irrigated. They further said that at present only 10 lakh acres are under irrigation and after restoration of the tanks 10 lakh acres more could be irrigated.

However, there is large difference between the above quoted figures to the government published statistics – year-wise. In Telangana 11.10 lakh acres were irrigated in 1955-56 and this was reduced to 5.70 lakh acres in 1999-00 – see the book published by the author in 2000 with the title "Andhra Pradesh Agriculture: Scenario of the last four decades". This means, the government quoted figures are more by around 40% of the real area. There is a need to relook in to the figures presented by the Telangana government. The benefit that is going to be achieved after restoration of the tanks will be around 15-20% only at the cost of Rs. 25,000 crores. This is the reality.

Like the proverb "the causes for Mahabharata legendry Karna's death were innumerable", the causes for the reduction in area under irrigation from 1955-56 to 1999-00 to present day in new Andhra Pradesh and Telangana states were also innumerable. Among them the primary cause is due to tanks encroachments; and the second one is the silt formation. For example in the district of Hyderabad there used to be 932 tanks covering an area of 55,000 acres in which 65-70 of the area was encroached and the rest become cesspool of poison with pollution. The reduction in irrigated area is not only confined to new Andhra Pradesh and Telangana but it is a common feature at India – World over. In the case of India, the area under irrigation in 1960-61 was 114.0 lakh acres and the same was 73.5 lakh acres in 1999-00. In the last seven years [2008-2014] under different government schemes de-silting was under taken three times in several tanks.

Water Grid Project: To provide drinking water to households in Telangana, known as "Water Grid", government announced allocation of Rs. 40,000 crores without going in to the details on this subject matter. Now media reports that government is ready to call for tenders under 26 packages costing Rs. 34,568 crores and of which in the first phase consists of 11 packages costing Rs.

15,987 crores. Government has fixed the pipe size v rate and payment schedule. On paper it is O.K. but at this massive scale how the government is going to achieve the quality does a big question under the corrupt regime exist in the state. Few days back a TV channel presented a story on the pipelines quality – wherein the material fixed on the pipes has been gone. Here one major issue is the availability of water on sustainable basis and allocated water form part of tribunal award or from other source should be clearly specified. Otherwise they become disputed water and create problems. They must link existing water grids in different districts.

When the Successive governments, including the present government, is unable to execute the drinking water supply to the households in the capital city Hyderabad, can the government execute such a massive water grid scheme? Within Hyderabad around 380 MGD of water [excluding water available through groundwater] the government has account for 45% of this capacity only. Of the remaining 55% of water is lost through pilferage [60%] and through leakages [40%]. The leakages are causing severe water pollution problems. In fact I requested the Chief Secretary of Telangana State to bring out a "White Paper" on the expenditure incurred by the government on these projects and what will be the new expenditure vis-vis water grid and Kakatiya Mission cost and deduction of money under national project to Chevella-Pranahita, etc. The government has a fundamental duty to present the facts other-wise it will burden them in future. It must also include what will happen if a new government forms after 2019 elections? Government is asking farmers to wait two more years to get power and same way the farming will face drought conditions for more years.

2.6 Concluding Remarks

Ancient rulers built water resources and our modern rulers destroying them and polluting them. The traditional politicians put importance in building irrigation projects but the modern regional political party politicians are looking at how much I can cash from a project even if it goes against the people's interests. Even the investigating agencies that bring out lapses are working on political lines. This is leading poor becoming poorer and rich becoming richer and wealth is going in to few hands.

The Central and State Governments must develop a system to utilize the wastewater generated in urban areas before they are dumped in to water bodies or rivers. Violators should be punishable on two counts – fine and jail terms; State governments are looking at bring water from faraway places by spending thousands of crores but they are least interested in protecting the local drinking water resources;

In the case of lift irrigation schemes including wells and bore-wells, governments must make it mandatory to use micro-irrigation practices and crop rotation with less water intensive crops/ cropping systems. Central government must look in to wasteful expenditure that burdens subsequent governments must put under check. Day by day the groundwater availability is going down as governments are not looking at ways and means of replenishing the groundwater.

Inter-state Irrigation Projects – Role of Tribunals

3.1 Introduction

Interstate projects move three steps forward and six steps backward due to several teething problems. The major ones are "tribunals and disputes". This chapter deals with the role of tribunals in resolving interstate water sharing issues. As an example Krishna River water distribution issue is discussed. The next chapter deals with the role of disputes in interstate irrigation projects. As an example Polavaram Project on Godavari River issue is discussed. These two chapters deal the main contentious issues of interstate irrigation projects in the development and use of water.

In resolving the water sharing issue of Krishna River among the three riparian states, namely Maharashtra, Karnataka and Andhra Pradesh [AP] -- until and otherwise specified, it refers to united Andhra Pradesh only [undivided], Justice Bachawat Tribunal [**KWDT-I**] was appointed by the Government of India in April 1969. 1976 May 27 the tribunal submitted its award/order to the Government of India. However, KWDT-I suggested in their order that if necessary Government of India can appoint a second tribunal to look in to the matter after 2000 May 31. Government of India appointed Justice Brijesh Kumar Tribunal [**KWDT-II**] on 2004 April 2. The tribunal started its work on 2006 August and submitted its first order on 2010 December 30 & the final order on November 30, 2013 to the Government of India. With the unfettered powers through "technical fraud" presented the report. In this chapter these are discussed.

However, AP Government approached the Supreme Court of India on KWDT-II award and on this Hon'ble court has granted stay in publishing the award in Gazette of India. Still the case is in the court. The tribunal [KWDT-II] consists of three retired judges, namely Justice Brijesh Kumar, former judge of Supreme Court of India; Justice S. P. Srivastava, former judge of Allahabad High Court, UP; and Justice D. K. Shaw, former judge, Calcutta High Court, Kolkata – now one of them is died. In his place a retired Judge from Karnataka was proposed by the Chief Justice of Supreme Court, who himself belongs to Karnataka. This was opposed by AP in the Supreme Court hearings. Most unfortunate thing is, UPA government extended the KWDT-II term for resolving the issue of sharing water among the two newly formed states from AP, namely "AP & TS" even though AP is disputing the award presented by them. This is a political game played by Karnataka politicians in UPA government at that time – in fact they are behind the appointment of the tribunal members. Also, they are the main players in the bifurcation of AP to serve their vested interests.

Krishna River: Krishna River originates in the Western Ghats near Mahabaleshwaram at an elevation of about 1,337 meter, in the state of Maharashtra about 64 km from the Arabian Sea in central India. It is one of the longest rivers in India. It is around 1,290 km in length and the fourth longest river which flows entirely in India, after the Ganga, Godavari and Narmada. It flows through the states of Maharashtra, Karnataka and Andhra Pradesh [AP] – now Telangana [TS] & Andhra Pradesh [AP] before outfalls in to Bay of Bengal at Hamasaladeevi in AP. The principal tributaries joining Krishna are the Ghataprabha, the Malaprabha, the Bhima, the Tungabhadra and the Musi as well the Koyna, the Yerla, the Warna, the Dindi and the Dudhganga. The largest tributary of the Krishna River is the Tungabhadra with a drainage basin measuring 71,417 km^2 running for about 531 km but the longest tributary is the Bhima River which makes a total run of 861 km and has an equally large drainage area of 70,614 km^2.

Most of this basin comprises rolling and undulating country, except for the western border, which is formed by an unbroken line of the Western Ghats. The river basin is approximately 200 meter deep. Krishna Basin extends over an area of 258,948 km^2 which is nearly 8% of the total geographical area of the country. This large basin lies in the states of Karnataka (113,210 km^2), TS & AP (76,252 km^2) and Maharashtra (69,425 km^2). The important soil types found in the basin are black soils, red soils, lateritic soils, alluvium, mixed soils, and saline and alkaline soils. Mullayanagiri peak, in Karnataka, is the highest point (1,930 m) of the Krishna basin. There are many dams constructed across the Krishna River, namely Almatti, Amar, Dhom, Jurala, Kanur, Nagarjuna Sagar, Narayanpur, Prakasham Barrage, Pulichinthala & Srisailam.

River Krishna is dying at an increasing rate. The river receives the waste from the large number of cities. Algal bloom in reservoirs, high alkalinity of river water, cyanide pollution from gold mines, no effective flood control plan, Alkali salts / high pH ash water runoff from coal fired power stations, inadequate salt export to sea leading to formation of saline/alkaline soils, excessive exploitation of river water causing insufficient environmental flows, astral land erosion due to inadequate water reaching the Sea, excessive silting of reservoirs due to deforestation and mining activities and Poor reservoirs management in terms of irrigation water supply, power generation & flood control. It causes heavy soil erosion during the monsoon season. During this time, the Krishna takes fertile soil from Maharashtra, Karnataka and TS towards the delta region of AP. It flows fast and furious, often reaching depths of over 75 feet (23 m). 2009 floods present an example to this.

3.2 Historical Perspectives of Tribunals

Water institutions have their origins in pre-independence legislation. The history of institutional development started with GoI Act 1919. The GoI Act 1935 drew attention explicitly to river disputes between one province and another. Sections 130 to 134 of this Act include water. The next stage is the constitution of India 1950 in which Articles 239 to 242 were worded on the same lines as sections 130 to 134 of the 1935 Act. Seventh amendment introduced the Article 262 "Adjudication of disputes relating to waters of inter-state rivers or river valleys", which leads us to what water laws and federal water bodies in India stands today.

According to Article 262: (1) Parliament may by law provide for the adjudication of any dispute or complaint with respect to the use, distribution or control of the waters of, or in, any inter-State river or river valley. Article 262: (2) Notwithstanding anything in this Constitution, Parliament may by law provide that neither the Supreme Court nor any other court shall exercise jurisdiction in respect of any such dispute or complaint as is referred to in clause (1). Provisions in Seventh Schedule: (a) *"List I - Union List" (Entry 56)* – "Regulation and development of Inter-State Rivers and river valleys to the extent to which such regulation and development under the control of the Union declared by law to be expedient in the public interest"; (b)*"List II - State List" (Entry 17)* – "Water, that is to say, water supplies, irrigation and canals, drainage an embankments, water storage and water power subject to the provisions of List I"; and (c)*"List III - Concurrent List" (Entry 20)* – There is no entry on water but there is an entry on planning, under "Economic and Social Planning". Since water is a significant input in agricultural development and industrial development, which are indicators of economic development, and since water is a primary need (drinking and sanitation) for social planning, water resource development could be covered under Concurrent List also. Only Entry 17 of List II has been in operation all along. However, Entry 20 of List III (Concurrent List) could be also said to have operated indirectly in view of the fact that the Central Government, through the Planning Commission, has to clear Water Resources Development projects for investments if these projects are to be eligible for central funds.

Article 263 of the Indian Constitution envisages establishing an Inter-State Council (ISC) with the mandate of enquiring into and advising upon disputes arising between the various states of India, to investigate subjects of common interest amongst the states, and to make recommendations upon such subjects for the better coordination of policy and action. The ISCI was finally established by presidential order on 28 May 1990 as a recommendatory body to fulfill the already mentioned constitutional mandate. The council comprises of the prime minister of India; chief ministers of all states; chief ministers of union territories; administrators of union territories; six ministers of cabinet rank in the union council of ministers and permanent invitees. Any matter in the Union list, Concurrent list or the state list of the Constitution of India in respect of which there exists a common interest as referred to in clause (a) of paragraph iv of the said order or a need for better coordination as referred to in clause (b) of the paragraph can be considered.

Legislative outcomes under Article 262 of the Constitution, Parliament has enacted the Inter-State Water Disputes Act (1956). This Act is to provide for the adjudication of disputes relating to waters of Inter-State Rivers and River Valleys. The Act came into effect on 28 August 1956, has been modified from time to time, and was last amended on 18 March 2002 Section 14, to achieve the objectives set forth. When any request is received from the state government in respect of any water dispute and the central government is of the opinion that the water dispute cannot be settled by negotiations, the central government is empowered to constitute a water disputes tribunal for the adjudication of the dispute by notifying in the official gazette.

The tribunal thus set up then has to investigate the matters referred to it and forward a report setting out the facts found by it and giving its decision on the same within a period of three years. The above Act has been used to set up several Tribunals to settle the Inter-State Water disputes. Under Entry 56 of List I, Parliament has enacted the River Boards Act (1956). The first act made provisions for setting up of river boards or advisory bodies by the central government at the request of the interested parties. These boards were to have two functions: 1) They would help to bring about proper and optimum utilization of the water resources of inter-state rivers; and 2) They would promote and operate schemes for irrigation, water supply, drainage, development of hydroelectric power and flood control.

National Water Policy, 1987: The broad objective of the guidelines governing the allocation of water is defined as "developing the waters of Inter-State River for the betterment of the population of the co-basin States/Union Territories to the extent such developments are not detrimental to the interests of other co-basin States". This national water policy of 1987 was amended in 2002 wherein Section 21 of this water policy deals with distribution of water amongst the states.

Conflicts and Settlement Mechanism: The major causes of conflicts in river water sharing can be grouped in two categories, technical and non-technical. Technical: differences in the approach for planning, design, construction, and operation of joint projects on Trans-boundary Rivers; Disagreement on the basic hydrological data and the actual present utilization of water; and Lack of openness and transparency in the exchange of data and information. Non-Technical: Violation of agreements by one party or the other; Disagreement on the basis and modalities of water sharing; Disagreement on riparian rights and basis thereof; the political system of India is based on multi-party democracy. Every political party gives a top slot to water resources development in its election manifesto.

Principles used for Settlement: Many options at dispute settlement have been employed, some with excellent results and some with continuing resentment and legal battles in the courts. (a) The Helsinki Rules11 on Equitable distribution and now the UN Law on Non Navigational Uses of International Water Courses are widely referred to. Even in the case of International Treaties and Agreements and these principles are followed; (b) The Principles of equitable distribution of water availability assessed at agreed locations on the main river and or its tributaries (on 75% dependability or average availability) have been followed in the Inter-State water disputes; (c) Changes of State boundaries due to the reorganization of States have brought even past agreements in dispute. This has resulted in establishing Tribunals to settle the sharing arrangements; (d) Allocation of water utilization for non-riparian states: Non-riparian States have been allocated water for utilization for drinking water, irrigation and other beneficial use. Helsinki rules are the best-known attempt to formulate principles for equitable allocation in the context of international water disputes. The International Law Association adopted these in 1966 at Helsinki. These rules extend up to 37 articles. Articles 4 and 5 cover procedures for preventing and settling disputes and according to article 4, "each basin is entitled, within its territory, to a reasonable and equitable share in the beneficial use of water of an international drainage basin" and Article 5 sets out 11 factors, which will determine what is reasonable and equitable share. The 11 factors are: 1. The geography of the basin, including the extent of the drainage area in the territory of each basin state; 2. The hydrology of the basin, including the contribution of water by each basin states; 3. The climate affecting the basin; 4. The economic and social needs of each basin state; 5. The population dependant on water of each basin state; 6. The comparative costs of alternative means of satisfying the economic and social needs of each basin state; 7. The availability of other resources; 8. The avoidance of unnecessary waste in the utilization of waters of the basin 9. The practicability of compensation to one or more of the co-basin states as a means of adjusting conflicts among uses; 10. The degree to which each basin State may be satisfied without causing substantial injury to a co-basin; 11. The past utilization of the waters of the basin, in particular existing utilization; uses on the basis of Agreements of the riparian States considering the established water shortages and hardship in such States or towns or cities.

The Settlement Mechanism: (a) The negotiation route: Over 130 Agreements have been evolved on the sharing of Interstate River waters or on specific projects. All these agreements have used the negotiation route, with the Central Government playing the pivotal role under the Constitutional Laws, Acts, and Statutory Rules. Most of these Agreements have worked well since they were done with the willing consent of the Party States to the Inter-State Basin. (b) Settlements through treaties: It is increasingly being recognized that maintaining a certain minimum flow in the rivers during the lean season months for ecological considerations is necessary, and provisions have been made for the same in the new agreement (Upper Yamuna) and treaty (Mahakali Treaty) signed in recent years. (c) Interstate river water projects funded by the central government: Since most of the river basins of India are Inter-State in character, the Central Organizations viz., the Planning Commission [now it is replaced by Neeti Ayog a political body] and the Ministry of Water Resources with its technical attached organization, the Central Water Commission, have exercised a very well set schedule of techno-economic clearance guidelines in approving the Inter-State projects planned by the States for implementation under the Five Year Plans. This procedure has been institutionalized, even though it is time consuming. This route of clearance ensures that projects on the Inter-State Rivers are not taken up without an agreement on water sharing in general, or project specific sharing in particular, of the waters of the river basin. There is a loophole in this, since the clearance is required only if the State wants Central Plan funding for the project. Otherwise, the State can go ahead with the project if funds are not a constraint. In that case, the aggrieved States can seek judicial intervention to stop the project.

India's Experience: The Inter-State Water Disputes Act seems to provide fairly clear procedures for handling disputes. At the same time, however, "the law permits considerable discretion, and different disputes have followed quite different paths to settlement", or in a few cases, continued disagreement. The central government has given substantial attention to water disputes, which began to emerge soon after the framing of the Constitution. As far back as 1967, the following 15 cases were identified, divided into two groups. The first group was those cases where interstate agreements through mutual discussions and negotiations had been successfully reached: 1. Musakhand Project dispute between Uttar Pradesh and Bihar, settled in 1964; 2. Tungabhadra Project High-level canal dispute between Karnataka and Andhra Pradesh, settled in 1956; 3. Sharing of costs and benefits of Jamni Dam Project between Uttar Pradesh and Madhya Pradesh, settled in 1965 ; 4. Palar water dispute between Tamil Nadu and Karnataka, settled in 1956; 5. Sharing of Subarnarekha river water among Bihar, Orissa and West Bengal, settled in 1964; 6. Exploitation of Mahi river water between Gujarat and Rajasthan, settled in 1966; 7. Utilization of Ravi-Beas waters between Punjab, Rajasthan, Jammu and Kashmir, settled in 1965.

The second group discussed consists of those cases, which had not been settled at that time: 1. The Krishna - Godavari waters dispute among Maharashtra, Karnataka, Andhra Pradesh and Orissa; 2. The Cauvery water dispute among Tamil Nadu, Karnataka and Kerala; 3. The Narmada water dispute among Gujarat, MP, Maharashtra and Rajasthan; 4. The Tungabhadra project issues other than the high level canal between Karnataka and Andhra Pradesh; 5. The issue of extension of irrigation from the Rangwan Dam of UP between UP and MP; 6. The Koymani river dispute between Bihar and West Bengal; 7. The dispute over the Keolari Nadi waters between MP and UP; 8. The Bandar Canal project, affecting Madhya Pradesh and Uttar Pradesh; A study of the details of these cases clearly puts them in two groups. The first three on this list were or are major disputes, involving large river basins. They were all ultimately referred to tribunals, with varying degrees of success. The last five cases on the list are actually closer in characteristics (relatively small and specific) to the most of the cases on the first list. Three cases which involve important disputes, and illustrate well the variety of paths that disputes have taken in the Indian institutional context, namely (a) The Krishna-Godavari water dispute, (b) The Cauvery Water Disputes Tribunal (c) The Sutlej Water Dispute.

Enforcement of Tribunal Award: This issue was given some attention by the Sarkaria Commission. It noted that section 6 of the ISWD act of 1956 provides that the Union Government shall publish the decision of the Tribunal in the Official Gazette and the decision shall be final and binding on the parties to the dispute and shall be given effect by them (Government of India, 1988, Chapter 17.4.18, p. 491). The commission's report goes on to suggest that the center cannot enforce the tribunal award if a state government refuses to implement the award. It notes that the amendment of the act in 1980, inserting section 6A, which provides for an agency to implement a tribunal award, is not sufficient because such an agency cannot function without the cooperation of the states concerned. The Sarkaria Commission's recommendation is, therefore, that a water tribunal's award should have the same force and sanction behind it as an order or decree of the Supreme Court. We recommend that the Act should be suitably amended for this purpose. (Government of India, 1988, Chapter 17.4.19, p. 491) but it should be noted that water tribunals already have such court equivalent powers for a narrow range of issues, including gathering of information, requiring witnesses to testify, and recovering the costs of the tribunal (Section 9 of the ISWD Act). Furthermore, the ISWD Act, Section 11 states that notwithstanding anything contained in any other law, neither the Supreme Court nor any other court shall have or exercise jurisdiction in respect of any water dispute which may be referred to a Tribunal under this Act. One possible interpretation of this provision is that it does implicitly give water tribunals broadly an equivalent status to the Supreme Court, and their decisions must have the same force.

Once again, the resolution of water disputes is complicated by being tangled in the general difficulties of center-state federal issues. Thus the recommendation to amend the act might not get to the crux of the problem. The Sarkaria Commission's other recommendations were based on the same kinds of difficulties in resolving past disputes. Two recommendations related to placing time limits on constituting tribunals and having them deliver decisions. These merely echoed the recommendations of the Administrative Reforms Commission (1969, Chapter V) nearly 20 years before. Another recommendation was that the center could appoint a tribunal without being asked to do so by a state government. A final recommendation was for the establishment of a national level data bank and information system. None of these recommendations has been carried out.

Summary: Water is precious natural resource and no nation can afford to ignore the imperative need, for comprehensive planning to secure optimum utilization of its water resources. The Irrigation Commission had recommended the establishment of a National Water Resources Council, as a policy making apex body with adequate technical support.

National Water Resources Council, which was constituted in 1983, met for the first time in October 1985. In this meeting, the Council was unanimous that water should be treated as a precious scarce national resource and dealt with as such, and that there was urgent need for formulation of a national water policy with a view to ensuring optimum use of available water resources, both surface and ground, in national interest. In its second meeting held on September 9, 1987, the Council adopted a National Water Policy. This policy emphasizes, among other things, that planning and development of water resources, also including inter-basin transfer of water, should be governed by a national perspective.

The National Water Resources Council in its meeting in October 1985 has recognized the need for the establishment of a Data Bank and an information system at the national level: "We fully endorse the need for such a Data Bank and information system and recommend that adequate machinery should be set up for this purpose at the earliest. We also recommend that there should be a provision in the Act that States shall be required to give necessary data for which purpose the Tribunal shall be vested with powers of a Court."

The National Commission on Agriculture observed that "in view of the inadequacy of water resources to meet the future agricultural and other requirements in many parts of the country, it becomes a matter of great national importance to conserve and utilize them most judiciously and economically". Comprehensive plans for the inter-State rivers covering irrigation, drainage, drinking water, inland water-ways and hydroelectric power generation should be prepared. Preparation of such plans requires proper evaluation of all existing water resources. This reinforces the need for proper institutional infrastructure for this purpose. The National Commission on Agriculture also observed: "A river basin, and in the case of large rivers a sub-basin, is a natural unit for such a plan, as it has a defined watershed boundary and within it there is an inter-relationship between the surface and ground water resources. The river basin plan should present a comprehensive outline of the development possibilities of the land and water resources of the basin, establish priorities in respect of water use for various purposes, indicate the need for earmarking water for any specific purpose and indicate priority of projects".

All the above presented theoretical exercises are on paper and look excellent but they serve very little in practical sense. That is there is wide gap between theory and practice with loopholes and flexibilities in those theoretical exercises. It is most unfortunate that the Sarkaria Commission report looked at giving more teeth to Tribunals and their award. They did not anticipated the otherwise, namely if tribunal give patrician reports in favour of a riparian state as they are invariably appointed through the law ministry. This is clear evident from the Krishna tribunal and its award, which is discussed below.

3.3   Issues Pertaining to Data Selection

3.3.1 Data Selection

In river water sharing among riparian states or among regions in a state primarily depends upon the water availability data series built from that river catchment area over years. In the case of KWDT-I & KWDT-II appointed by the Central Government on Krishna River, used different data series and presented widely differing awards.

At the time of writing the report KWDT-I used all the 78 years [1894-95 to 1971-72] data that was available at that time. This was agreed by the three riparian states. KWDT-I observed that the volume of water which passes over and through Vijayawada weir would give us a fair idea of the volume of flow in the river after the upstream utilizations are added to it.

At the time of writing the report KWDT-II has 114 years [1884-85 to 2007-08] data series but used only a part of this data of 47 years [1961-62 to 2007-08]. AP disagreed on such selection as this represents the period of high rainfall. KWDT-II rejected the data used by the KWDT-I. KWDT-II put forth some subjective, unscientific and illogical arguments for selecting the 47 years data instead of 114 years data that was available to them for use; and for rejecting 78 years data series used by KWDT-I. They are given as follows:

- On KWDT-I data series, KWDT-II observed that "It is simply commendable. It has been very rightly agreed by the parties". On the page 246 [page numbers refer to those seen in online report of KWDT-II] of the report it states that "They do not match hence cannot be integrated";

- On page 250 KWDT-II states that "The longer the time series, however, greater the chance that it is neither stationary, consistent, nor homogeneous. The later part of long time series can present a better data set if it is reasonable to expect that similar condition will prevail in future";

- On page 274 KWDT-II states that "We are of the opinion that 47 years length of a series should be considered sufficient to assess water availability of river. It more than fulfills the minimum requirements of IS Code ---";

- On page 304 KWDT-II states that "such increase as reflected seems to be quite natural & obvious. The utilization has more than doubled since 1971-72.The increase is therefore clearly seems to be on account of the return flows and addition of increase in storage and utilization in minor irrigation".

These issues are discussed in the next section. Though KWDT-II selected 47 years data series but used five widely differing data sets including parts of 78 years of KWDT-I data series which was rejected by them to prove their pre-conceived objective. They are:

- Used 78 years [1894-95 to 1971-72] data to define water availability at 75% probability level – Scheme A of KWDT-I;

- Used 47 years [1961-62 to 2007-08] data to define values at 65% probability level and the Mean [meets at 58% probability level] – this is not truly Scheme B of KWDT-I;

- Used 26 years data [1981-82 to 2007-08] of Krishna-Bhima River zones to raise the Almatti Dam height;

- Used one year [2006-07] data of water use to change the dependable water availability from world widely used 75% level to 65% level;

- Used data from 1941-42 to 2007-08 to prove delta is getting its share of water even under low availability of water and thus wants show that the 78 years data set is not accurate.

3.3.2   Dependable Water and Water Allocations

KWDT-I divided water distribution among the riparian states in to Scheme A and Scheme B. For Scheme A, KWDT-I defined "dependable" water. Internationally accepted level of probability to define "dependability" is 75% probability level. The irrigation commission recommended to continuing the practice to design supplies at 75% dependability for irrigation schemes. 75% probability is being widely [nationally & internationally, including India Meteorological Department, ICRISAT Hyderabad, etc] used as dependable – which means available in three out of four years. KWDT-I also followed this and used water available at 75% probability level as dependable – though AP proposed higher probability level.

For estimating the dependable water, KWDT-I adopted very simple graphical technique [crude method] by plotting the 78 years water availability data series in an ascending order [the lowest to the highest] against the probability levels 100% to 0.0% [**Figure 3.1**]. The value at 75% probability level in **Figure 3.1** is

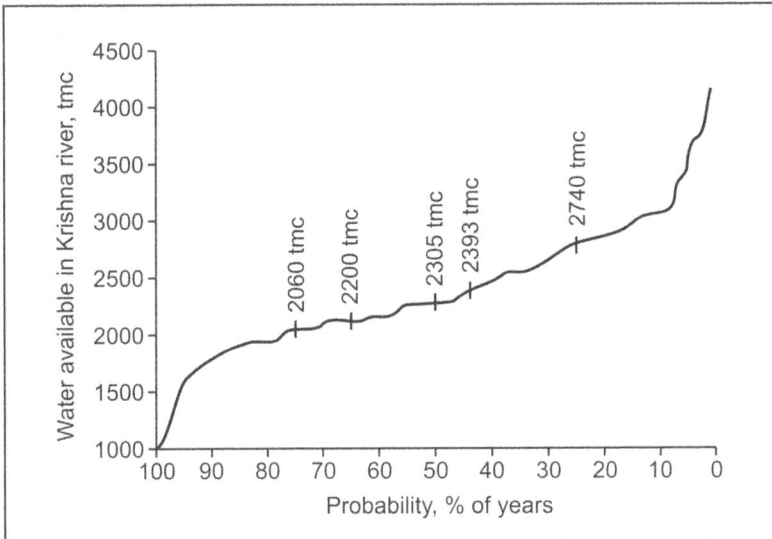

Figure 3.1: Probability curve of 78 years data series

2060 TMC. However, for distribution, KWDT-I added from return flows as 70 TMC. Therefore, KWDT-I distributed 2060 + 70 = 2130 TMC among the three riparian states [**Table 3.1**] under Scheme A.

Table 3.1: Krishna river water allocations among the three Riparian states

Water allocations,	TMC			
	Total	Maharashtra	Karnataka	Andhra Pradesh
(a) According to Justice Bachawat Tribunal – now this is in operation				
	2130*	585	734	811
(b) According to Justice Brijesh Kumar Tribunal				
	2578	666	911	1001 + 4 #
(i)	163	46	72	45
(ii)	285	35	105	145

* Water available at 75% probability level (2060 TMC) + return flow into river (70 TMC)

  (i)   additional water available at 65% probability level over 75% probability level [2293 – 2130 TMC]

 (ii)   additional water available at mean [58% probability level] over 65% probability level [2578 – 2293 TMC]

KWDT-I further observed that "Scheme B sharing of surplus/deficit water as the case may be, each year was to be **shared by the three riparian states**. However, it will not be proper to set up any authority without consent of parties. In this connection, the government of India under clarification No. 6 pleaded before the tribunal that have not expressly provided for the sharing of the deficiency in the river flows when monsoon fail and drought conditions occur. KWDT-I emphasized that AP state will be at liberty to use the excess flow in surplus years and must bear the burden of the deficiency in the river Krishna in the lien years.

It means AP will get allocated water of 811 TMC three out of four years. To compensate one of the four years in which AP get less than 811 TMC, AP was allowed to use surplus water [if available] in the remaining 75% surplus years. That is, KWDT-I allowed AP to use the surplus water over 2130 and at the same time stated that AP also must bear the burden of the deficiency [less than 2130 TMC years] in the river flows in the 25% lien years.

However, AP did not built projects to utilize the surplus water until 2004 as this water being illegally utilized beyond Srisailam dam. N. T. Rama Rao [**NTR**] the then AP Chief Minister from TDP only laid foundation stones for 8 projects identified in dry regions of the state. Chandrababu Naidu [**CBN**] became Chief Minister by dethroning NTR again laid foundation stones where NTR laid foundation stones. At the same time Karnataka & Maharashtra built several projects illegally [still continued in building illegal projects], though they have no allocations from the surplus flows.

KWDT-II followed Scheme A of KWDT-I as per **Figure 3.1**. However, KWDT-II used 65% probability level as dependable. KWDT-II stated that the use of water during 2006-07 matches with 65% probability limit value and thus KWDT-II has chosen 65% as dependable. This is highly subjective judgment wherein one year comparison used to define the whole spectrum of data of very low to very high river flows to favour Karnataka and Maharashtra states that goes against KWDT-I order. Now where in the world used one year comparison to define river flow distribution among riparian states. KWDT-II with unfettered powers rejected the opposition. With this AP will get less than the allocated water in 35% of the years instead 25% of the years.

Figure 3.2 presents the probability curve built based on the 47 years data series [1961-62 to 2007-08] of KWDT-II similar to **Figure 3.1**. The value at 65% probability level from this figure is 2293 TMC. This is 163 TMC more than that of 2130 TMC. Also, KWDT-II indirectly agreed to implement the request by the Karnataka state that dependable probability may be raised to 50% probability level. KWDT-II as third stage distributed water between the Mean and the 65% probability level. The Mean from the **Figure 3.2** is 2578 TMC which is available at 58% probability level. Thus the difference is 285 [= 2578-2293] TMC. That means KWDT-II distributed 448 [= 163 + 285] TMC over 2130 TMC of Scheme A of KWDT-I to the three riparian states in two stages [**Table 3.1**]. In all the three stages, the upstream riparian states were given right to first use their allocated water.

KWDT-I observed that "We have, however, provided that the authority or the tribunal which will be reviewing the order of this tribunal shall not as far as practicable, disturb any utilization that may be undertaken by any state within the limits of the allocations made it by the tribunal". All the actions of KWDT-II go against this. KWDT-I allowed AP to use surplus water in 75% of the years [if available] to compensate the deficit in 25% of the years. Now, they changed to 42% of the years.

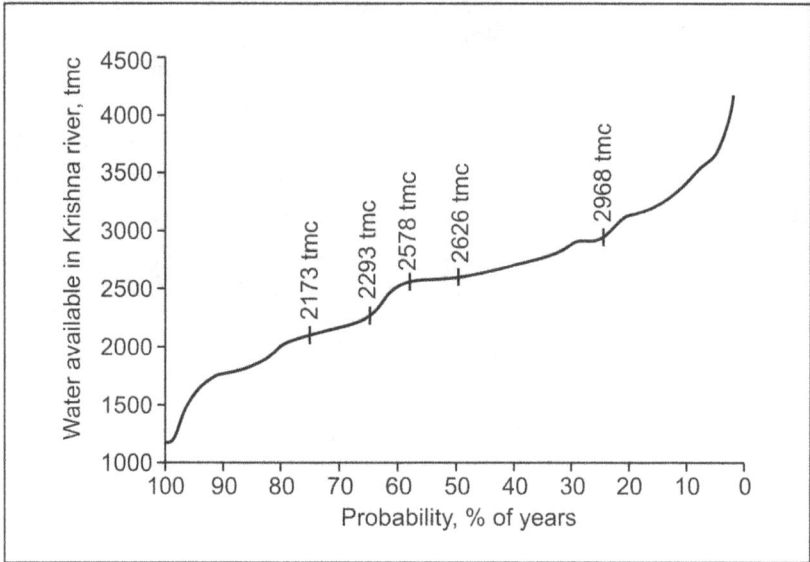

Figure 3.2: Probability curve of 47 years data series

By this, AP will have to bear the deficit at three stages, namely 75, 65 & 58% as AP get it share after the two upstream states fully utilize their share. As a result AP must bear the burden of the deficiency in the river Krishna in the lien years up to 42% of the years in three stages. However, the 65% & 58% probability levels will be further reduced by the "triple impact effect". This I termed as Technical Fraud". These in turn are going to affect the projects built/under construction by AP to utilize allocated water and surplus water use as per the KWDT-I. Here there is another important point that needs to be taken in to account. Due to cyclonic activity, there is high probability that the flood water between Nagarjunasagar and Prakasham barrage will automatically goes in to the Sea as the capacity of Prakasham barrage is small. This is not accounted while assessing the water availability in the Krishna River. In **Figures 3.1 & 3.2** the amount of water below 20% probability may not be truly available as seen in the figures. *KWDT-II followed illogical arguments while selecting the data series to define the water availability; with poor mathematical manipulation KWDT-II allowed Karnataka to raise the Almatti Dam height; and with bad argument KWDT-II allowed to convert illegal projects as legal in Karnataka.*

KWDT-I observed that "hereby declares that the state of Maharashtra, Karnataka and Andhra Pradesh will be free to make use of underground water within their respective state territories in the Krishna basin. The use of groundwater by any state was not to be reckoned as use of water of the river Krishna. Diversion of water of an inter-state river outside the river basin is legal for optimum utilization of water. However, these two have negative impacts on AP with KWDT-II order. KWDT-II nowhere in the report/award discussed all these direct and indirect impacts of such actions on AP water availability. These are discussed in the later sections of the chapter.

3.3.3   Illogical and Unscientific Arguments

KWDT-II put forth several illogical arguments to justify their actions with unfettered powers conferred to the tribunals in the constitution of the tribunals. Let us see them below one by one:

Continuity – Moving Average Technique & Data Requirement: In nature, data series follow different patterns, namely random, linear, non-linear, cyclic, etc. Therefore as a first step, KWDT-II should have analyzed the data series for its pattern of change with time before making the statement on continuity. Unfortunately, KWDT-II hasn't made any such analysis but simply made a statement that "they do not match hence cannot be integrated". This is inaccurate biased statement. Let us see this:

World Meteorological Organization of United Nations [WMO/UN] in 1966 [WMO, 1966] presented a manual on "Climate Change" to help to understand periodicity and trend in meteorological data series. This book is also available in India Meteorological Department/Pune Library. The issue was discussed in my book – **Reddy [1993].** For the identification of climate fluctuations in rainfall data one needs long and continuous rainfall records. The results to be reasonably valid the minimum data period required to adopt power spectrum analysis is twice that of expected periodicity. That is if the expected periodicity is 22 years then we need more than 44 years data series; if the expected periodicity is 60 years, then we need more than 120 years data series, etc.

When we don't know the expected periodicity we can use "Moving Average Technique" or "Iterative Auto-Regression Technique" to quantify it. Also, the acknowledged purpose of computing moving average of climate series is to smooth away the short term (rapid) period variations so that the longer (slower) period variations can be revealed more clearly. If there are well defined long period variations in the series this devise is quite useful in revealing their form. This procedure was used in the case of onset of southwest monsoon over Kerala Coast [**Reddy, 1977**] – see **Figure 3.3**.

Very recently British Royal Society and US National Academy of Sciences brought out an overview "Climate change: Evidence & causes" [24th February 2014]. The report included a figure of annual march of global temperature anomaly [Figure 4 on page 13] along with 10 year, 30 year & 60 year moving average patterns using 1850 to 2010 data series – see **Figure 3.4**. Here, the 60 year moving average pattern showed the trend after eliminating 60 year rhythmic variation.

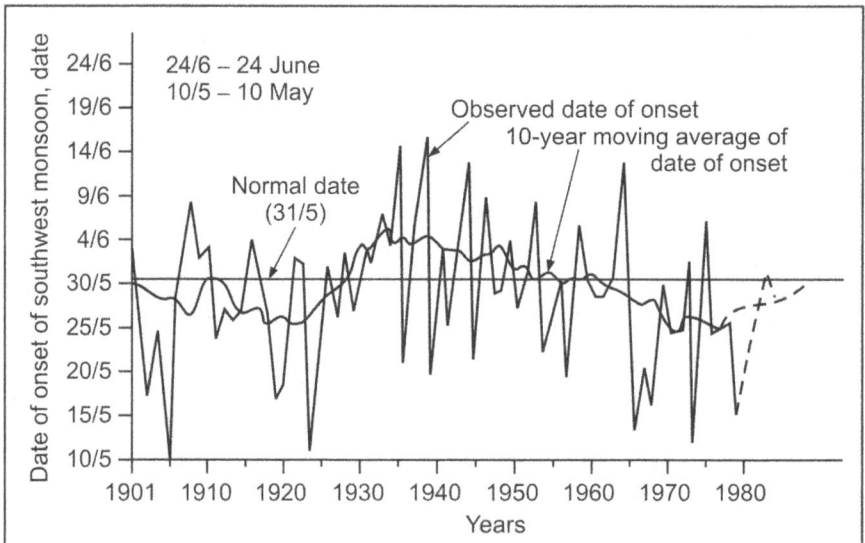

Figure 3.3: Time series of dates of onset of monsoon over Kerala coast

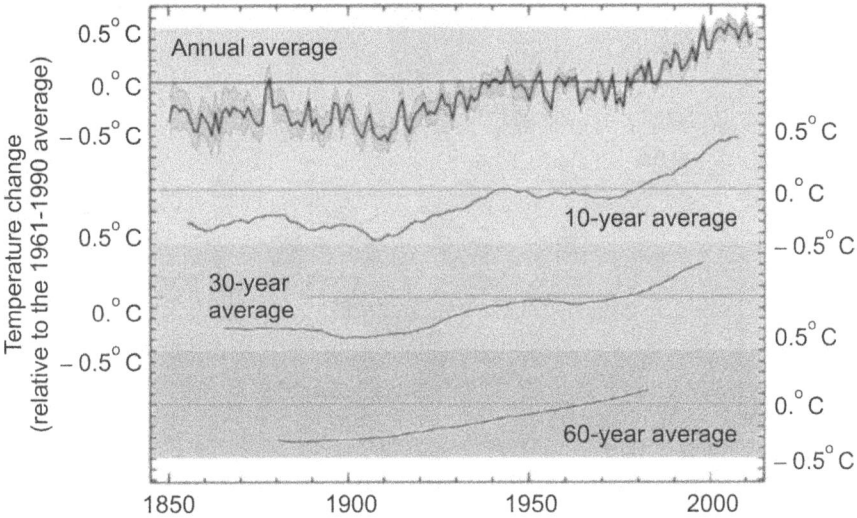

Figure 3.4: Importance of moving average

All these clearly reveal that moving average technique is a powerful tool to understand the rhythmic variation, if present, in climate data series. KWDT-II has not made any such attempt on its own.

Reddy (1993) presented the cyclic nature of precipitation data series and its application in dry-land agriculture planning. The Indian rainfall as well the rainfall of AP -- in the Krishna River catchment area -- present cyclic variations (**Reddy, 2008 & 2010**). Any analysis, more particularly probability analysis, using a truncated data of a cyclic variation data series gives misleading and biased results [**Reddy, 1993**]. It is exactly where the KWDT-II erred. **Figure 3.5** presents the annual march of southwest monsoon rainfall over India and **Figure 3.6** presents the annual march of annual precipitation of AP with reference to respective data series averages. The data presented in **Figures 3.5 & 3.6** was taken from **Parthasarathy, et al. (1995)** publication.

Figure 3.5 presents a 60-year cyclic variation – it follows a sine curve, 0.0 to +1.0 to 0.0 to -1.0 to 0.0. That is, if the first 30 years data series follow the above the average pattern [wherein flood years are more frequent than drought years], then the next 30 years follow the below the average pattern [wherein drought

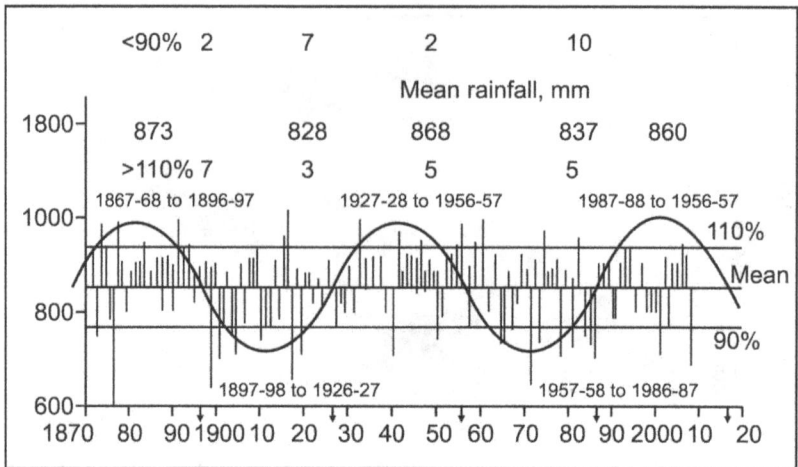

Figure 3.5: Annual march of All-India southwest monsoon rainfall

Table 3.2: Frequency of occurrence of floods in few selected northwest Indian Rivers

Frequency of high magnitude floods*			
River	Period	Frequency	Climatic cycle
Chenab	1962-1990	1 in 9 years	(a) Below the average cycle
	1990-1998	1 in 3 years	(b) Above the average cycle
Ravi	1963-1990	1 in 14 years	(c) Below the average cycle
	1990-1998	1 in 3 years	(d) Above the average cycle
Beas	1941-1990	1 in 8 years	(e) Below the average cycle
	1990-1995	1 in 2 years	(f) Above the average cycle
*State of Environment Report, India – 2009, MoEF/GoI: The frequency of floods in India is largely due to deforestation in the catchment area, destruction of surface vegetation, changes in land use, increased urbanization and other developmental activities – this is a false statement but it is more in association of cyclic variation in rainfall.			

years are more frequent than flood years] and vice-versa. This pattern repeats every 60 years like our 60-year astrological calendar. Here flood and drought years refer to years with rainfall amount more than 110% and less than 90% of the average rainfall, respectively. In **Figure 3.5,** already two 60-year cycles have been completed and third 60-year cycle started in 1987/88 and will continue up to around 2046/47. River flows also follow the rainfall pattern. This can be seen from **Table 3.2**.This table presents the frequency of occurrence of floods in rivers Chenab, Ravi & Beas, which showed a pattern similar to the 60-year cycle of all-India Southwest Monsoon rainfall over India   [*this data was taken from the MoEF/GoI publication "State of Environment Report, India, 2009"*].

AP and the adjoining Krishna River catchment area not only receive rains during the Southwest Monsoon season [June to September] but also during the Northeast Monsoon season [October to December]. Majority of severe cyclonic storms are confined to the Northeast Monsoon season. Also, pre-monsoon season [April & May] receive cyclones. Annual rainfall takes all these factors in to account – in the case of All India Southwest Monsoon [SWM] rainfall, around 78% of the annual rainfall is received on an average in the SWM alone. Because of these, the cyclic variation of this region presents a different pattern **[Figure 3.6]**. This figure presents 132 year cycle in which the first 66-years presented below the average pattern – observed **24 drought years and 12 flood years** – and the next 66-years presented above the average pattern – observed **24 flood years and 12 drought years** – completed by 2001; and next 66-years below the average pattern commenced in 2001.

Figure 3.7 presents the annual march of water available in Krishna River during 1894-95 to 2007-08 [114 years] – covering the data of both from KWDT-I & KWDT-II with the overlapping years of 1961-62 to 1971-72. The pattern of **Figure 3.7** follows exactly the pattern seen in **Figure 3.6**. It is natural as water availability in any river follow the precipitation pattern in the catchment area of that river. Thus, during below the average rainfall period the water availability was on lower side [below the average on majority of the years] and during above the average rainfall period the water availability was on higher side [above the average on majority of the years]. *Thus KWDT-II data is a continuum of the data of KWDT-I only.*

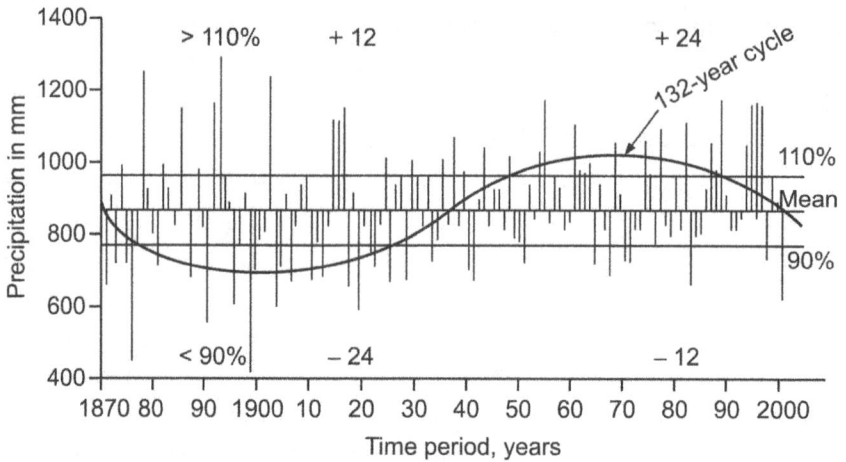

Figure 3.6: Annual march of annual rainfall of Andhra Pradesh

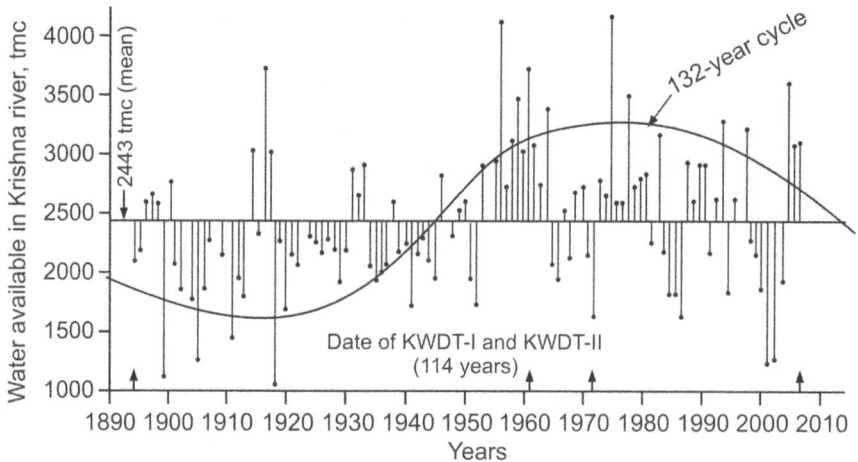

Figure 3.7: Annual march of annual water availability in Krishna river

Reasons attributed to high water availability: On pages 304-305 – KWDT-II argued that "It is to be noticed that in average flow there is an increase of about 180 TMC as compared to the series prepared before by KWDT-I and the series of 112 years; and at 75% dependability there is an increase of about 113 TMC. Such increase as reflected seems to be quite natural and obvious. The utilizations have more than doubled since 1971-72. The increase therefore clearly seems to be on account of the return flows and addition of increase in storages and utilization in minor irrigation. The availability of water thus has increased rather than decreased as has been tried to be shown on behalf of the State of AP". Also, KWDT-II on page 385 state that "Annexure-II to APAD-63 is not accepted as correct figures. We have considered the figures of gross flows and utilization as per chart prepared for 47 years' series to assess yield of the river Krishna".

KWDT-II has chosen 40 years of above the average part of the 132-year cycle 1961-2000 and 7 years of below the average cycle of 2001-2007 [**Figure 3.6**]. So, the data is heavily biased by the high rainfall period. The average of 78 years data series of KWDT-I is 2393 TMC and the same with 47 years data series of KWDT-II is 2578 TMC, in excess by 185 TMC over the KWDT-I mean. 2578 TMC is available at 34% probability level on KWDT-I [**Figure 3.1**] probability curve instead of 58% probability level [on **Figure 3.2**]. The mean of 114 years data series [**Figure 3.8** presents the probability curve] is 2443 TMC and the value 2578 TMC is available at 41.5% probability level [**Figure 3.8**] instead of 58% probability level.

Therefore to get meaningful and unbiased values of water availability at any given probability level, the data period selection plays crucial role when the data series follow a cyclic pattern. Whether they are short or long period, they must cover one full cycle or more cycles. The truncated data of a cycle gives misleading inferences. While all India Southwest Monsoon case we need 60 or multiples of 60 years data series and in the case of AP annual rainfall, we need 132 or multiples of 132 years data series to get unbiased probability estimates. KWDT-II used truncated data of a cycle and that too representing better rainfall period.

The criteria for classification of minor irrigation schemes have been changing from time to time. Since April 1993 all groundwater schemes and surface water schemes (both flow and lift) having cultivable command area up to 2000 hectares individually are considered as minor irrigation schemes. In the case of tank irrigation, the area under cultivation has come down by more than 50%. The details can be seen from **Reddy (2014)**: The area under tanks irrigation during 1960-61 was 11.51 Lha has come down to 6.51 Lha by 1999-00 in AP. Same is the case all over India, including Karnataka & Maharashtra, wherein 45.6 Mha has come down to 29.4 Mha during 1960-61 to 1999-2000. Thus, proportionately the return flows, if any, will decrease and will not increase. In the case of wells/bore-wells they increased from 3.25 Lha to 19.00 Lha in the case of AP; and from 7.29 Mha to 24.70 Mha at all India level. With drastic reduction in ground water availability area cultivated per pump in AP has come down from 2.5 acres to 0.5 acres. Here the return flows are insignificant. Also, well/bore-well water use does not come under water use classification of KWDT-I.

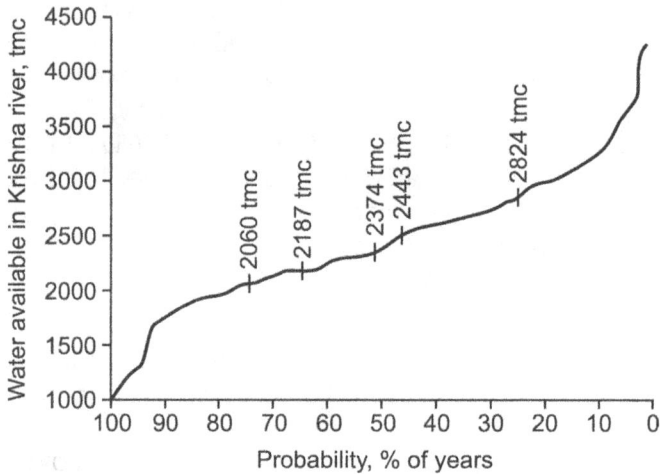

Figure 3.8: Probability curve of 114 years data series

It is pertinent to note that KWDT-II to substantiate another postulation used part of KWDT-I under Scheme A allocations and rejected under Scheme B allocations. KWDT-II used 1941-42 to 1968-69 data series from KWDT-I plus 1972-73 to 2007-08 data series from KWDT-II to show that delta areas of AP are receiving its share of allocated water of 181.20 TMC as a last point in the chain of Krishna water utilization. On pages 371-372, 379-384, etc., of the report of KWDT-II clearly indicated that there is nothing wrong with the data prior to 1961-62 and yet KWDT-II on page 385 of the report observed that "Annexure-II to APAD-63 is not accepted as correct figures. We have considered the figures of gross flows and utilization as per chart prepared for 47 years series to assess yield of the river Krishna".

In fact though in the earlier part of the data period all the projects in the upstream were not ready to use water. And yet, the withdrawals in delta area, a last point of withdrawal, showed lower values. This pattern clearly supports the fact that under low rainfall conditions, the water availability condition in AP will be expected very low under all projects in operation now.

KWDT-I submitted its report/order to Government of India on 1976 May 27, which mean that the Scheme A was not in operation before that date in many projects. That means by that date all projects in the upstream of Prakasham Barrage [the last downstream barrage on the river Krishna] were not in operation and yet delta water withdrawals were lower than 181.2 TMC [data presented on pages 379-384 of the KWDT-II] during the low water availability period – also it was a low rainfall period. Take the water use by delta, for example, during the period 1941-42 to 1961-62: 149, 155, 183, 164, 165, 186, 175, 179, 155, 178, 177, 161, 167, 156, 161, 147, 173, 157, 177 & 201 TMC. Out of these 20 years the delta used less than 181.2 TMC of allocated water in 17 years. The same during 1972-73 to 1991-92 [20 years] this was zero. If we look at the data of 1991-92 to 2005-06 they are 191, 181, 234, 237, 188, 192, 234, 224, 233, 221, 190, 118, 84, 137 & 187 TMC. Out of these 15 years, four years are less than 181.2 TMC. In the last four years water entered in to the Sea were given as 111, 13, 12 & 23 TMC. Also, these are consecutive years. That means the additional storage facility of 150 TMC has no meaning in the consecutive deficit years [like 2014 & 2015]. Here one more thing we need to keep in mind is that KWDT-I allowed to-use surplus water, if any, by dry areas but the Government has not created any such facility and thus excess water use in surplus years was reflected during 1972-73 to 1991-92 under good rainfall.

In 2006-07 the amount of water entered the sea was 1273 TMC out of 3624 TMC and delta used only 187.0 TMC; in 2000-01 they are respectively 365, 2185 & 221 TMC; in 2004-05 they are respectively 12, 1934 & 137 TMC. This shows high year to year variability in water entering the Sea and water used by delta. The higher values of entering the Sea must be carefully evaluated in terms of the cyclonic floods between Nagarjunasagar dam and Prakasham barrage.

With the changing rainfall pattern – year to year and long-term – and cyclonic activity the available water varies. The return flows and additional storage relates to water availability in any given year and the preceding year and that too distribution within the season and not on the average. So also, the evaporation component varies with the rainfall pattern and the year & season of occurrence of cyclones. In this particular case severe storms in the Northeast Monsoon season play an important role in the water going as waste in to the Sea.

Water availability is different from water utilization/success rate. When water is available under high rainfall conditions and projects are there to utilize the water, then utilization increases; and when there is less water with poor rainfall conditions the utilization will be low. Also, the storage and return flows follow the water availability and thus utilization. Therefore, increase is not due to return flows and storage and utilization in minor irrigation but because availability of more water due to high rainfall. This is quite obvious from the Scheme A of KWDT-I. This also results in the wastage of water, which enter the Sea finally. This is evident from the data presented in pages 302-303 & 379-384 relating flows in to the Sea, water withdrawals in delta area and total water availability after 1960-61 and water withdrawals in delta area prior to 1961-62.

From all these discussions, it is clear that the two statements mentioned above by KWDT-II are erroneous and not based on scientific analysis. It is quite obvious as KWDT-II is not technical experts.

Homogeneity and Data Length: In the literature there are several models to estimate the minimum expected amounts at a given probability level – **Figure 3.9,** copy of a page from the book of **Reddy (1993).** The accuracy of the estimates depends upon the degree of skewness in the data sets. In statistical terms, thus, to obtain accurate estimates the data series must follow normal distribution, wherein the Mean coincides with 50% probability level. The degree of skewness increases with the deviation of the Mean from 50% -- generally we call that data series as skewed, positively or negatively. When the Mean is at higher than 50% probability level then it is said that the data set is positively skewed and thus biased by higher values; and when the Mean is at lower than 50% probability level then it is said that the data series is negatively skewed and thus biased by lower values.

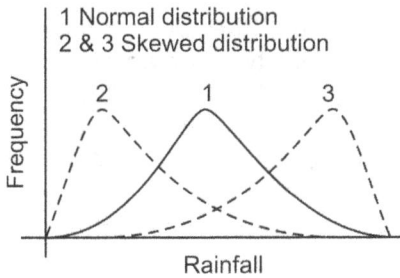

Figure 3.9: Normal distribution

It is seen that the Mean of KWDT-II is realized at 58% with probability curve showing sudden jumps – **Figure 3.2**, which means that the data series is highly positively skewed and thus highly non-homogeneous; the Mean of KWDT-I is realized at 43% – **Figure 3.1**, which means that the data series is negatively skewed and thus non-homogeneous and yet it is less than that of KWDT-II data series with no sudden jumps; the combined data series of 114 years the Mean is realized at 48% – **Figure 3.8**, which is very close to 50% and thus it is close to normal distribution and thus the data series are homogeneous. The combined data series of 114 years meets the condition "stationary, consistent, homogeneous" and on the contrary the data series of KWDT-II failed to meet these conditions. Next in order after 114 years data series is KWDT-I and not KWDT-II to meet these conditions. From this it is clear that the truncated data series of a cyclic pattern present positively or negatively skewed pattern based on the part from which the truncated data form part of the cyclic variation.

Therefore, KWDT-II erred by stating on pages 249-250 of the report that "The longer the time series, however, greater the chance that it is neither stationary, consistent, nor homogeneous". Here, it is important to note that the selected data series must be equally distributed on either side of the mean and not number of years or length of the data series. Also, it is important that the data series must cover full cycle [if the data series follow a cyclic pattern] or equally distributed on either side of the Mean of a normal curve pattern. The truncated data presenting skewed pattern gives misleading inferences. This is exactly what has happened with KWDT-II choice of 47 years data series, covering high rainfall period water availability.

Similar Conditions: Thus, the best series to be use for the estimation of water availability at different probability levels is 114 years, which was in fact available to KWDT-II at that time, as there is strong continuity under cyclic pattern and is homogeneous with mean close to 50% probability level. 114 years data showed that 78 years data & 47 years data can be integrated through the cyclic pattern. Cyclic pattern in rainfall repeats the cycle [**Figure 3.7**] and thus the chances of occurrence of similar conditions of 47 years data are bleak in continuation. The data after 2000 showed the lower side of the cyclic variation – below the average pattern.

Summary: To understand, 'whether there is a continuity or not' in the rejected data of KWDT-I, it is important to know first 'what type of pattern' the data series form part. This was not assessed by KWDT-II in its voluminous report. The Indian rainfall as well the rainfall in the Krishna River catchment area presents cyclic variation. Any analysis, more particularly probability analysis, using a truncated data of a cycle variation data series gives misleading and biased results. It is exactly where KWDT-II erred. These issues in fact discussed the present author in his book [**Reddy, 1993**] using the data of different countries.

To get unbiased estimates at different probabilities, the data series must cover both the low and high rainfall periods when the data series follow a cyclic variation. 47 years data series cover principally high rainfall period. 114 years data series cover both the low and the high rainfall periods. Next in order comes 78 years data series, which used all available data at that time also to a certain extent covers both the low and the high rainfall periods.

Thus, the best series to use for the estimation of water availability at different probability levels is 114 years data series, which was in fact available to KWDT-II at that time, as there is strong continuity under cyclic pattern and is homogeneous as the data series being normal. 114-years data series showed that KWDT-I & KWDT-II data series could be integrated through the cyclic pattern. Cyclic pattern in rainfall repeats the cycle and thus the chances of occurrence of similar conditions of KWDT-II are bleak in continuation.

One can see large differences in the start [above 90%] and end [below 10%] parts of probability curves in **Figure 3.1, 3.2 & 3.8**. This is very important in deciding water allocations.

Therefore, KWDT-II erred by saying on page 246 of the report that "They do not match hence cannot be integrated"; on pages 249-250 "The later part of long time series can present a better data set if it is reasonable to expect that similar condition will prevail in future"; & on page 274 of the report states that "We are of the opinion that 47 years length of a series should be considered sufficient to assess water availability of a river. It more than fulfills the minimum requirements of IS Code ---". As the present data series follow a Sine Curve pattern, all the three observations are erroneous not based on science.

3.4 Drought Proneness and its Association with the Climate Change

On pages 660-661, while permitting allocation of water to some projects in Karnataka KWDT-II argued "Therefore, such grounds of the upper and lower riparian States [that means Maharashtra and AP], if allowed, would act as veto to topple the project UKP-III, which is to serve the needy drought prone areas in Karnataka lying within the basin of river Krishna".

In fact drought prone areas are not only confined to Karnataka but also this zone spreads from AP to Maharashtra via Karnataka, known as rain shadow zone. It is clear from this that KWDT-II put false argument to favour Karnataka. This can be seen from **Figure 3.10 [Reddy, 1993]**, wherein it presents the drought proneness map of India. From **Figure 3.10**, it is seen that under the rain shadow zone drought proneness ranges between 30 and 60% of the years on an average. However, with the cyclic variations in precipitation, the drought proneness varies between high to low around the mean with reference to high and low rainfall periods of the cycle. **Figure 3.11** presents the importance of climate change in terms of cycles in precipitation on drought condition using the data of Kurnool. On an average the drought risk is 45% of the years in Kurnool region of AP but it increased to 70% of the years during below the average cycle pattern period and reduced to 30% of the years during the above the average cycle pattern period. This issue was discussed by **Reddy [1993]** for different countries.

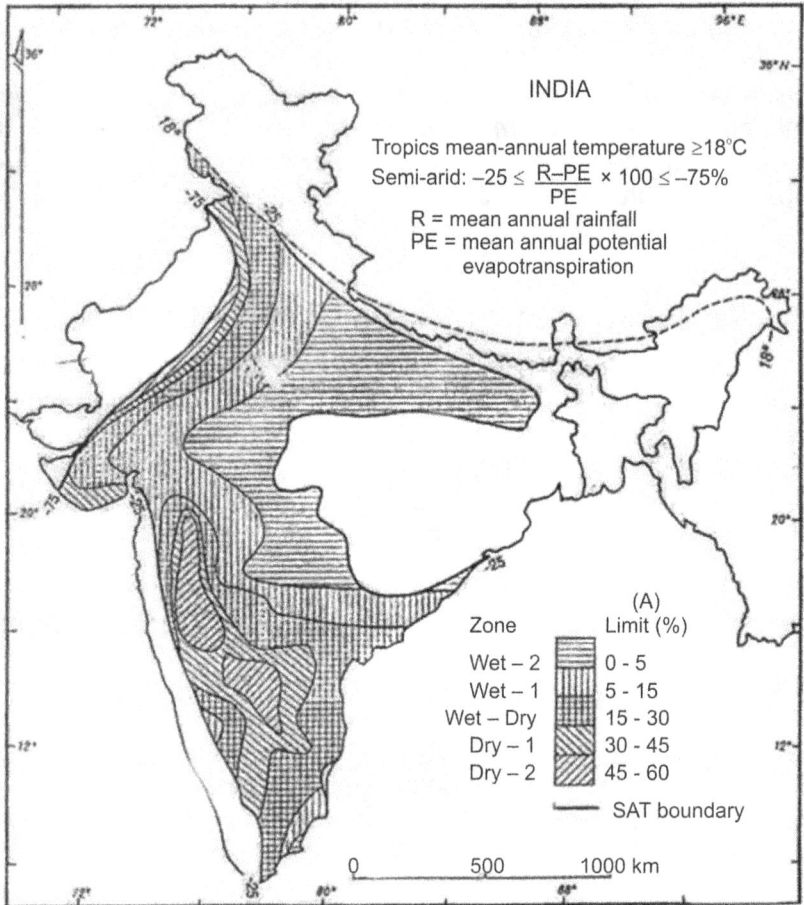

INDIA

Tropics mean-annual temperature $\geq 18°C$

Semi-arid: $-25 \leq \dfrac{R-PE}{PE} \times 100 \leq -75\%$

R = mean annual rainfall
PE = mean annual potential evapotranspiration

Zone	(A) Limit (%)
Wet – 2	0 - 5
Wet – 1	5 - 15
Wet – Dry	15 - 30
Dry – 1	30 - 45
Dry – 2	45 - 60
—— SAT boundary	

0   500   1000 km

Figure 3.10: Drought proneness map of India

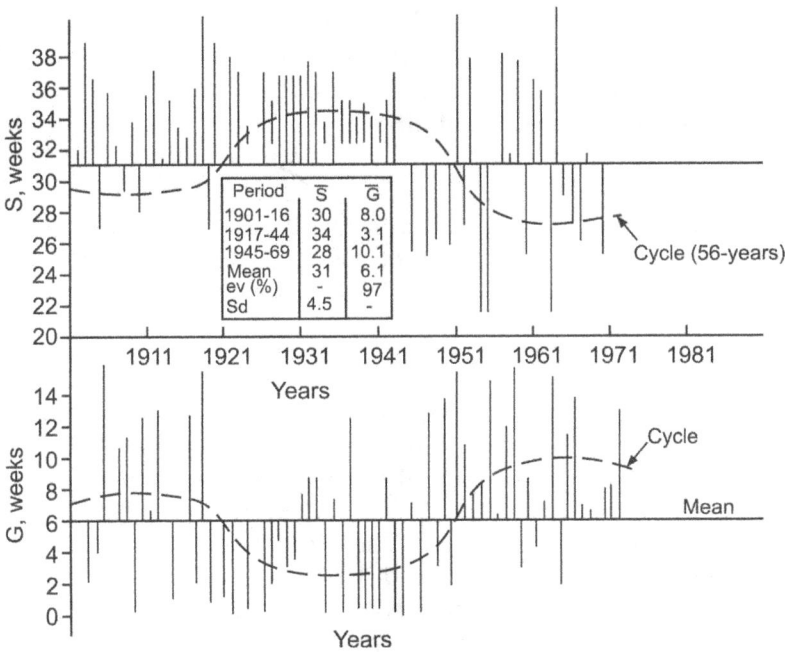

Figure 3.11: Drought under climate change conditions

From the above it is clear that the intension of KWDT-II is not to help drought prone areas in river Krishna basin but only to help Karnataka state at the cost of the other two riparian states, namely AP and Maharashtra under the disguise of drought proneness. With this KWDT-II legalized illegal projects and allocated 40 TMC of additional water.

AP started building projects to use surplus water as permitted by KWDT-I. KWDT-II did not care to allocate water to these projects on the same logic of drought proneness. Without any allocations by KWDT-I, Karnataka built several illegal projects and KWDT-II legalized them by allocating water under the disguise of drought prone areas. Even though the same is applicable to AP, no such allocations were made.

3.5   Raising of Almatti Dam Height to 524.256 m

3.5.1 A Manipulated Mathematics

Here we must note the fact that AP gets its share of water flows from Karnataka through Tungabhadra and Krishna Rivers.

Tungabhadra meets Krishna River after Kurnool Town and before Srisailam Dam. AP also receives some water from rains in the state associated with the monsoons and cyclonic activity. KWDT-II put forth a manipulated mathematics to show that availability of water to AP will not be affected by raising the Almatti Dam height from 519.6 m to 524.256 m. --Even the 519.6 m raise from the original height [515.0 m] was carried out when Dev Gouda from Karnataka was Prime Minister of India. -- The water availability manipulation is carried out by KWDT-II as follows:

KWDT-II instead using the 47 years data series, for this KWDT-II used 26 years (1981-82 to 2006-07) data related to Krishna-Bhima Rivers. The mean water flows in to AP based on the 26 years data series is 932 TMC. KWDT-II also stated that AP will get 190 TMC on an average from Tungabhadra and AP get locally 350-400 TMC on an average. With this, KWDT-II showed that even without raising the Almatti Dam height to 524.256 m, AP will be getting on an average around 1530 [= 932 + 190 + 350-400] TMC. With these calculations KWDT-II tried to show that even with the raising of Almatti Dam height to 524.256 m AP will be still getting 1300 [= 1530 − 230] TMC, which is far higher than the allocated 800 TMC by KWDT-I award/order!!! But the fact is, it is not the reality and the truth is hidden under a wrap. Let us see this manipulation: at what probability level this availability of water is achieved?

By adding water allocations to Karnataka & Maharashtra under KWDT-I to 1530 TMC, it becomes mean available water as 2830 TMC {= 1530 TMC, available in Andhra Pradesh + **1300** TMC [= 560 TMC, allocations to Maharashtra + 700 TMC, allocations to Karnataka + 40 TMC, additional allocations made by present tribunal to three projects under Tungabhadra part to Karnataka]}. Here, the return flows are not included [70 TMC = 25 + 34 + 11]. This is far higher than even the mean from the 47 years data series [2578 TMC] by 252 TMC; and under 78 years data series mean [2393 TMC] by 397 TMC. These were not discussed anywhere in the report.

If the allocated water by KWDT-II to Karnataka and Maharashtra [see **Table 3.1**] is used then 2830 TMC will go up to 3088 TMC. This is a major data manipulation initiated by the KWDT-II to prove their objective to raise the height of Almatti Dam to store additional water of 230 TMC and provide 40 TMC to three projects on Tungabhadra zone. Let us see these in a simple step-wise presentation:

Step 1:   KWDT-II data shows AP is getting on an average 1530 TMC without raising Almatti Dam height to 524.256 m;

Step 2:   KWDT-II allocated 40 TMC to Karnataka through four projects in Tungabhadra zone – some reports say 38 TMC;

Step 3:   KWDT-II allocated 230 TMC to Karnataka through rising Almatti Dam height – some reports say it is 130 TC;

Step 4:   KWDT-II argued that *even if 230 TMC of additional water is allocated* to Almatti Dam, AP will be getting 1300 [1530 – 230] TMC

Step 5:   Water allocated under KWDT-I at 75% probability level excluding return flows to Maharashtra [560 TMC] + Karnataka [700 TMC] is 1260 [= 560 + 700] TMC

Step 6:   Extra water allocated by KWDT-II over KWDT-I to Maharashtra [81 TMC] + Karnataka [177 TMC] is 258 TMC

Step 7:   Therefore the average water available under KWDT-I is: Step 1 plus Step 2 plus Step 5 = 2830 [= 1530 + 40 + 1260] TMC – this value is expected on probability curves given in **Figures 3.1, 3.2 & 3.8** respectively are: 21, 30 & 25 % probability levels;

Step 8:   Therefore the average water available under KWDT-II water allocations is Step 7 + Step 6 = 3088 [= 2830 + 258] TMC -- this value is expected on probability curves given in **Figures 3.1, 3.2 & 3.8** respectively are: 09, 27 & 14 % probability levels;

Step 9:   Probability levels given under Step 7 & Step 8 are further modified under two conditions, namely (a) With the raising of Almatti Dam height, KWDT-II allowing to use 230 TMC of water in addition to the existing level allocations. This will help increase in water spread area. This scenario will help the evaporation and infiltration in to groundwater from that water spread area, (b) With the cyclonic activity the major part of surplus water in many years goes into the sea – that is in **Figures 3.1, 3.2 & 3.8** the steep rise at the low probability side [that is below 20% probability levels] will not really available to AP or for that matter to Karnataka or Maharashtra;

Step 10:   Some reports show 38 in place of 40 TMC and 130 in place of 230 TMC [see steps – 1 & 2]. Even if it is the case, under KWDT-II water allocations, this is 2986 [3088 – 100 - 2] TMC and yet it is far higher than 2578 TMC [Mean]. In **Figure 3.8** it meets at around 15% probability level instead at 58% probability level. That means, out of 100 years only in 15 years that water is available. *That means, if KWDT-II award is implemented with Step 7, Step 8 and Step 9 conditions AP will not get even a single year its allocated water of 1005 TMC.*

Step 11:   In this assessment the 150 TMC of carryover storage to compensate deficit in 25% of the years is not included. This is presented later.

3.5.2 Practical Scenario

In the above presented analysis, KWDT-II made casually passing statement on low available water in some years but not studied its real impact on Andhra Pradesh. The mean of five consecutive low water available years [2000-01 to 2004-05] is 380 TMC, which is 40.8% of the mean 932 TMC. By adding another set of four consecutive years [1984-85-1987-88] the average of these two sets of data series [in all 9 years] comes as 494.3 TMC, is 53.0% of 932 TMC. This reduction is not confined to water entering in to AP from Krishna-Bhima Rivers' part but also to the total available water of 2830 TMC [or around 3088 TMC according to the case]. Under these two situations, the overall water availability comes down as: 2830 TMC × 40.8% = 1154.6 TMC; and 2830 × 53.0% = 1499.9 TMC. This shows that these values are far less than 2060 TMC [value at 75% probability level under Scheme A of KWDT-I, which is at present in operation]. Here the basic flaw is that KWDT-II hasn't taken into account the surplus water entering in to the Sea, more particularly during heavy rainfall years associated with cyclonic activity in to Bay of Bengal or Arabian Sea. As per KWDT-II award AP will be getting its share from Karnataka after Karnataka and Maharashtra uses their share of allocations. The share of combined allocations at 75% probability level is 1319 TMC, at 65% probability level it is 1437 TMC, and at mean [at 58% probability level] it is 1577 TMC. That mean AP will be getting practically nil flows from Karnataka and Maharashtra during the 9 years under KWDT-II award. 2015 is a classical example occurrence of very low water availability.

Here it is important to note the fact that AP will get less 800 TMC in 25% of the years and to compensate this deficiency KWDT-I allowed carryover storage of 150 TMC to AP. However, in consecutive years with deficit, this carryover has little use. Because of that only KWDT-I allowed using surplus water during surplus years by AP in 75% of the years. Now it comes down to less than or equal to 25% years only – this will go down to less than 20% under the average of 3088 TMC. This is going to create water wars among three riparian states, more particularly among the two bifurcated states of AP, namely AP & TS. The fraudulent tribunal [KWDT-II] was asked to look in to sharing of allocated water by the bifurcated states of AP. God alone knows what will happen with such a tribunal.

3.6 Technical Fraud in Krishna River Water Sharing

3.6.1 Triple Impact Effect

The triple impact effect as part of technical fraud enacted by KWDT-II is as follows:

First, through the data manipulation the tribunal raised the mean available water in the river Krishna for distribution among the three riparian states, namely Maharashtra, Karnataka and Andhra Pradesh [AP]. This is a direct impact;

Second, this raise in water availability, indirectly changed the probability level at which this water is available to AP. This is an indirect impact;

Third, through the data manipulation the tribunal allowed Karnataka to raise the Almatti Dam height. This is going to affect the probability at which AP is going to get its share of water. This is an indirect compounded impact.

KWDT-II nowhere discussed these direct and indirect impacts on water availability to AP, which is on the downstream of Krishna basin. The details of these three impacts are given as follows [**Table 3.3**]:

Table 3.3: Triple impact effect at a glance

Condition	Data sets				
	47 years	78 years		114 years	
	[1]	[2]	[1]-[2]	[3]	[1]-[3]
[a] Amount [in TMC] of available water realized for the condition [% probability]					
65%	2293	2150	143	2187	106
58%	2578	2200	378	2250	328
Mean	2578	2393	185	2443	135
50%	2626	2305	321	2374	252
[b] % probability at which the condition [in TMC] is realized					
2293	65	50		55	
2578	58	34		41.5	
Mean	58	43		48	
2830	30	21		25	
3088	27	09		14	

Direct Impact: KWDT-II inflated the water availability through the selection of high rainfall period for the distribution and these are presented in **Table 3.3.** Water realized values at 65% and Mean are taken from **Figures 3.1, 3.2 & 3.8** under three different data series. From **Table 3.3** it is clear that those KWDT-II values are higher by 93 TMC and 185 TMC at the 65% probability level and the Mean, respectively over KWDT-I values. The same under the 114 years data series they are higher by 106 TMC and 135 TMC, respectively.

Indirect Impact: KWDT-II adapted the same water availability value at 75% probability level as derived by KWDT-I using 78 years data series – this is also same under 114 data series. This can be seen from **Figure 3.1.** This is 2130 = [2060 + 70] TMC. **Figure 3.2** presents the probability curve based on the 47 years data series of KWDT-II. From this figure the water availability values at the 65% probability level and at the Mean [58% probability level] are 2293 TMC and 2578 TMC. These values, namely 2293 TMC & 2578 TMC are achieved in probability curve of KWDT-I [**Figure 3.1**] at 50% and 34% probability levels, respectively; and in probability curve of 114 years data series [**Figure 3.8**] they are achieved at 55 and 41.5% probability levels. These are presented in **Table 3.3**.

Indirect Compounded Impact: KWDT-II allowed raising Almatti Dam height to 524.256 m to store additional 230 TMC in Karnataka. KWDT-II also provided 40 TMC to projects under Tungabhadra zone in Karnataka. The total additional water provided to Karnataka comes around 270 [= 230 + 40] TMC. By this manipulation, the mean 2578 TMC in KWDT-II shifts to 2848 = 2578 + 270 TMC. The 2848 TMC, which is achieved at around 30%, 21% and 25%, respectively in **Figures 3.2, 3.1 & 3.8** and thus, 58% is reduced to 30, 21 & 25% levels. **Table 3.5** presents these results.

Table 3.4: Triple impact effect – A practical example

Year	Amount of water TMC*					
	[A]	[B]	[C]	[D]	[E]	[F]
1999-00	365	202	-083	095	233	2305
2000-01	252	089	-196	-018	221	2185
2001-02	111	-052	-337	-159	190	1836
2002-03	013	-150	-435	-257	118	1239
2003-04	012	-151	-436	-258	084	1252
2004-05	023	-140	-425	-247	137	1934
2005-06	1273	1110	825	+1003	187	3624
2006-07	944	781	496	+674	254	3186
2007-08	927	764	479	+657	235	3230
2008-09	502	339	054	+232		
*A = Excess or deficit over 2130 TMC; B = Excess or deficit over 2293 TMC [= A – 163]; C = Excess or deficit over 2578 TMC [= A – 448]; D = A - 270 TMC [= Almatti allocations of 230 TMC plus 40 TMC allocations to three projects under Tungabhadra]; E = water utilization in Delta [allocated 181.20 TMC] – Data from KWDT-II pages 379 to 384; F = Water available [part of 47 years KWDT-II data series]						

According to data manipulation figures presented in the above section, the means 2830 TMC and 3088 TMC are respectively with Karnataka and Maharashtra using water as per KWDT-I and KWDT-II. The probability levels at which these two values meet the probability curves in **Figures 5.1, 3.2 & 3.8** are presented in

Table 3.3. They are around 30, 21 & 25 and 27, 09 & 14 %, respectively. **Even if take the 270 TMC as 168 [130 + 38] TMC as reported by some reports that is discussed in the above section, this comes around 2986 TMC. It comes within 2830 and 3088 TMC only.**

This action automatically goes against the award given by the KWDT-I saying "we have, however, provided that the authority or the tribunal which will be reviewing the order of this tribunal shall not as far as practicable, disturb any utilization that may be undertaken by any state within the limits of the allocation made it by the tribunal". The KWDT-I also observed on for the sharing of the deficiency in the river flows when monsoon fail and drought conditions occur that "Andhra Pradesh state will be at liberty to use the excess flow in surplus years and must bear the burden of the deficiency in the river flows in the lien years". Thus, the triple impact effect resulted: AP used to get less than the allocated water in 25% of the years from the KWDT-I award but with the KWDT-II award AP will be getting less than the allocated water in more than 75% of the years due to triple impact effect. This is a technical fraud to favour Karnataka. Thus it negates the use of 150 TMC as storage allowed meeting the needs of AP during deficit years as the surplus years will be reduced from 75% to less than 25% of the years.

In this assessment the 150 TMC of storage to compensate deficit in 25% of the years is not included. This is presented later.

3.6.2 Important Note

It is a clear manipulation to favour Karnataka. KWDT-II argued that the 230 TMC is only used for power production and thus it goes back in to the river and will not affect the flows in to AP. This is only hypothetical. Karnataka has been going on building illegal projects nobody stopping them. Once it is legalized as per the KWDT-II Award, nobody can stop misuse of 230 TMC [if not 130 TMC] by Karnataka. Thus, it becomes practically a gift to Karnataka and thus it is a boon to Karnataka and bane to AP. Andhra Pradesh has to depend on the mercy of Karnataka.

We have seen the Maharashtra case wherein KWDT-I allowed to use water in hydropower production in another basin. This water is going as waste in to Arabian Sea after power production.

Also, the fact is when the Almatti Dam height is raised from 519.600 m to 524.256 m the water spread area increases multifold level and thus becomes a zone of groundwater recharging. This also increases the evaporation losses from the reservoir component. These issues were not discussed, on how much water is going as groundwater and how much is going as evaporation in such a scenario.

3.6.3 Practical Example to Understand "Triple Impact"

To understand the "Triple Impact" effects on AP's water availability, a case study is presented in **Table 3.4** using ten years [1999-00 to 2008-09] data taken from pages 302-303 & pages 379-384 of the KWDT-II report. It is clear that under KWDT-I Scheme A, all the ten years there was a flow in to the Sea [Column A in the table] irrespective of, weather the Delta area meeting its share or not, as Prakasham Barrage capacity is very small, even when Nagarjunasagar Reservoir is empty, water may go as wastewater in to Bay of Bengal with local flows, mainly associated with cyclonic activity between Nagarjunasagar Dam and Prakasham barrage, which was discussed earlier.

It is seen from column B in the table that Andhra Pradesh will receive less than the allocated 856 TMC [**Table 3.1**] at 65% probability level in 40% of the years and at mean [column C] AP receives allocated 1001+4 TMC [**Table 3.1**] of water in four out of the remaining six years. In all, in six years out of ten years AP receives less than the allocated water of 1005 TMC under KWDT-II. Then what will happen under poor monsoon years or jointly poor and good monsoon years put together – similar to 114 years data series? They will be reduced further.

Then, what will be the water availability to AP after deducting the 270 TMC from the water flowing in to the Sea under KWDT-II? From Column E, it is clear that out of ten years in five years there is no flow in to the Sea. That is, in 50% of the years AP will not get its allocation of 811 TMC leaving aside the 1005 TMC allocations of KWDT-II. It is pertinent to note that the observation of KWDT-1 that "We have, however, provided that authority or the Tribunal which will be reviewing the order of this tribunal shall not as for as practicable, disturb any utilization that may be undertaken by any state within the limits of the allocation made it by the tribunal

(KWDT-I)". However, KWDT-II executed exactly opposite to this by allowing Karnataka to use 270 TMC in excess over KWDT-I allocation. The storage facility of 150 TMC given to AP by KWDT-I has no utility during 2001-02 to 2004-05 consecutive years.

During 2005-06, though water availability was 3624 TMC [Column G], the water use in delta of AP was only 187 TMC [Column F] and water entered in to the Sea was 1273 TMC [Column E] – against the allocation of 181.20 TMC. This is the character of the rainfall of the region. Sudden rainfall in late season goes to the Sea as waste. September/October 2009 floods present the classical example of devastation in Andhra Pradesh with large part going in to the Sea as waste. Also, it is seen from the **Table 3.6** that even in years with delta area receiving the less than allocated water of 181.20 TMC [during 2002-03 to 2004-05] and yet there was a flow in to the Sea. Also, previous year's surplus stored in reservoirs or the return flows have not helped the delta area to meet its allocation in poor rainfall years.

With the thermal power plants are coming up in Karnataka and Maharashtra that require huge quantity of water, what will happen to water flowing down to AP?

KWDT-II hasn't bothered to look into: What will be the impact on their biased decision on AP's water availability. This example clearly reflects the bias in the KWDT-II verdict. Such biased report should be rejected out rightly and take stringent action against the members of the tribunal for giving such fraudulent verdict to favour Karnataka state at the cost of people of AP.

3.7 Final Stroke: A Very Important Component

AP started building projects based on the surplus water as provided by KWDT-I. After bifurcation of the state of AP in to AP and Telangana governments are going ahead with projects left and right without getting clarity on the availability of surplus water under different conditions that are prevailing at present.

Under KWDT-I Surplus Water Availability: KWDT-I provided a provision to AP that to compensate the deficit in 25% of the years, 150 TMC of water can be stored in surplus water years. That means the surplus water to the projects built on surplus water is available only when the water flows in Krishna River exceed 2210

[= 2060 + 150] TMC. On **Figure 3.1**, the value of 2210 is available at 65% probability level – this is same in **Figure 3.8**. That means AP will be getting in around 35% of the years "zero" surplus water.

Under KWDT-II Surplus Water Availability: In the case of KWDT-II the value of 2210 is replaced by 2728 [= 2578 + 150] TMC. In **Figure 3.2** the value of 2728 TMC is available at 38% probability level – this is at 28% in **Figure 3.8**. That means in 62% of the years under 47 years data series or 72% of years under 114 years data series there will be "zero" surplus water available to AP.

With Almatti Dam Height 524.256 m Surplus Water Availability: Under KWDT-II, if we add the water allocations to Almatti dam [168 or 270 TMC] the value 2728 changes to 2998 [= 2728 + 270] or 2896 [= 2728 + 168] TMC. These values are available in **Figure 3.2** at 24% probability level or 30% probability level – the same in **Figure 3.8** are 17% or 19%.

By taking other Factors Affecting Surplus Water Availability: By taking other factors affecting surplus water availability referred under 2.5 and elsewhere under KWDT-II, award surplus water will not be available not even a single year to projects built on the surplus water by AP. KWDT-II hasn't discussed these issues.

3.8   Concluding Remarks

About the ground water use: KWDT-I allowed free use of ground water in the respective states and said this will not come under the water use. While writing this recommendation, KWDT-I might not have envisaged the present scenario, namely in upstream states expanding reservoirs. This is the case with increased area under water spread in Almatti Dam with 230 [or 130] TMC of additional storage facility and 40 [or 38] TMC under Tungabhadra zone. This facilitate excess water component under evaporation from the reservoirs and seepage to groundwater. These were not accounted by KWDT-II in the hundreds of pages report/award.

Diversion of Water: KWDT-I allowed diversion of an inter-state river outside the river basin for optimum utilization of water. The danger in this is wastage of water. Here the main issue is Maharashtra state diverting part of its allocated water to hydropower production in the outside the river basin and releasing the water in to Arabian Sea after power production. This means water is wasted to that extent instead of making this water available to areas in downstream of the Krishna Basin.

Change in Probability Levels: With the KWDT-II order, if it is implemented, the water to the projects proposed based on surplus water availability in Andhra Pradesh will be available only once in four or five years or none instead of three out of four years. These are presented results as summary with 47 years data series scenario and 114 years data scenario.

KWDT-II to favour Karnataka at the cost of Andhra Pradesh followed several steps of fraudulent path. Under the 47 years data series probability curve of KWDT-II, the six paths of water availability to AP are given as follows:

- **Step 1:** Andhra Pradesh will get less than the allocated water in 25% of the years at 2130 TMC;
- **Step 2:** Andhra Pradesh will get less than the allocated water in 35% of the years at 2293 TMC;
- **Step 3:** Andhra Pradesh will get less than the allocated water in 42% of the years at 2578 TMC;
- **Step 4:** Andhra Pradesh will get less than the allocated water in 70% of the years at 2830/2848 TMC;
- **Step 5:** Andhra Pradesh will get less than the allocated water in 73% of the years at 3088 TMC
- **Step 6:** Steps 4 & Step 5 will be modified in the 0 to 20% probability ranges in probability curves as the surplus water mainly goes in to the Sea – AP will get less than the allocated water in 100% of the years depending up on this scenario!!!

Under the 114 years data series probability curve – which is the real situation under Krishna basin, the six paths of water availability to AP are given as follows:

- **Step 1:** Andhra Pradesh will get less than the allocated water in 25% of the years at 2130 TMC;
- **Step 2:** Andhra Pradesh will get less than the allocated water in 45% of the years at 2293 TMC;
- **Step 3:** Andhra Pradesh will get less than the allocated water in 58.5 of the years at 2578 TMC;

- **Step 4:** Andhra Pradesh will get less than the allocated water in 75% of the years at 2830/2848 TMC;
- **Step 5:** Andhra Pradesh will get less than the allocated water in 86% of the years at 3088 TMC
- **Step 6:** Steps 4 & Step 5 will be modified in the 0 to 20% probability ranges in probability curves the surplus water mainly goes in to the Sea – AP will get less than the allocated water in 100% of the years depending up on the scenario!!!

Surplus Water Estimation is Erroneous: The surplus waters in many cases are associated with cyclonic activity and floods occur between Nagarjunasagar dam and Prakasham barrage, thus the projects will not get even that water as presented above. In the **Figure 3.8**, the steep rise below 10% probability levels is the classical example [similar to steep fall above 90% probability level. One is representing drought condition and the other severe flood conditions. Therefore, while assessing the water availability this factor must be taken in to account through the reduction associated with cyclonic rains.

Therefore, it is clear that when we do some acts, we must look at both at direct and indirect impacts of such propositions; and must take in to spectrum of issues associated with such phenomenon. In fact this is part of EIA notifications [1994 & 2006], wherein we look in to direct, indirect and compounded impacts on environment. This is not happening basically because of political interferences in technical decisions.

Projects Built on Surplus Water Availability: There are no chances that AP will get surplus water to projects built under surplus water use, even in a single year under KWDT-II. KWDT-II failed to discuss these important issues.

Inter-state Irrigation Projects – Disputes

4.1 Introduction

All most all irrigation projects that were/are taken up by different states in the country to meet their water needs are invariably suffering from either interstate or intrastate disputes or political tangle. Unfortunately different departments of Central Government are acting as agents of ruling junta. They instead solving the issues make them more complicated and make them to continue with infighting by spending crores of rupees as legal fee and at the same time wasting courts time. It is Indian culture. This goes like two monkeys and a cat with one roti sharing proverbial story.

The laws are weak and ambiguous; and they are changed as per the needs of ruling junta. Central Government introduced "EIA Notification, 1994" to get environmental clearance to projects under different categories. With the passing of time several amendments were proposed and in 2006 this was replaced by "EIA Notification, 2006". Again this is also corrupted with several amendments. The present NDA government even without any thought removed Planning Commission and in its place proposed a political body known as "Neeti-Ayog"; and weakened forest acts and environmental acts as it is being an anti-environment and a pro-industry government. This is also reflected the way the government [Finance Ministry & RBI] bring down interest rates to benefit business interests even after banks highlighting the non-returning of loan money to the tune of few lakh crores; and lowering the interest rates severely affect around 80% of the citizens who live on the interests. Another example is NDA government's proposed amendments to Land Act of 2013, though finally withdrew with strong opposition from other political parties. This wasted parliament valuable time and further put a hitch saying that individual states can promulgate their own Land Laws!!! This highlights the poor quality governance prevailing in Indian democracy.

Indian water-dispute settlement mechanisms are ambiguous and opaque. Nevertheless, water disputes are sometimes settled. There are plenty of reports, which tried to compare the disputes and suggested different mechanisms existing in the finance sector as a Guide and as well internationally known mechanisms. Unfortunately, we are forgetting the fact that again they depended on political set up and have their loopholes more badly than the water disputes redressed. Also the existing institutional frame work to "carry out the water balance and other studies...for optimum utilization of water resources..." like National Water Development Agency, 1992. This agency is a Government of India Society in the Ministry of Water Resources, and not a body with any statutory backing. Furthermore, its scope is technical, and separate from the institutional realities of water allocation. In 1983, the National Water Resources Council (NWRC) was created by a central government resolution. Its composition includes chief ministers of states, lieutenant governors of union territories, several central government ministers, and the prime minister as chairman. This group met first in October 1985, and adopted a National Water Policy in 1987. This policy emphasizes an integrated and environmentally sound basis for developing national water resources, but provides no specific recommendations for institutions to achieve this. Though the council was created out of disenchantment with the adjudicatory process for inter-state river disputes, it has not provided concrete proposals to improve that process, nor has it provided the useful alternative that was hoped for, as the persistence of the Ravi-Beas and Cauvery disputes indicates. It is like IPCC [Intergovernmental panel on Climate Change] that sub-serve the vested interests of political bosses. One thing we must keep in mind, when we talk of disputes, is that there are no parallels as each entity is separate by itself and we can write a separate book on each dispute, which has been mosqued by political vested interests. To understand this scenario, the case of Polavaram project on Godavari River is presented in this chapter.

Cross-border International Scenario: India, Pakistan, Bangladesh, Nepal, and Bhutan depend on rivers originating in and passing through foreign territory. China holds an advantageous position, as two most expansive river systems (the Brahmaputra and the Indus) originate from the glaciers in Chinese-controlled Tibetan plateau. The westward flowing Indus and Sutlej Rivers start in Tibet and pass through India and Pakistan, while the Brahmaputra traverses through China, India, and Bangladesh, aided by a couple of tributaries from Bhutan as well. India also shares several smaller rivers with Nepal and four other rivers with Pakistan. All of trans-boundary Rivers nourish India's lands on their forward journey to other countries, thus making it a significant player in all possible disputes. About one-third of its surface water is dependent on foreign-originating rivers.

Reports state that India has been a firm believer in negotiations both inside and outside the country. In fact, the archrivals have successfully honoured the Indus Water Treaty governing rights over the shared rivers. India has also actively engaged Bangladesh, Bhutan, and Nepal to settle disputes around rivers from time to time. China is the only country with which India has no agreement, except the one that improved sharing of data regarding river flow level during flood season. The lack of knowledge-sharing on the projects by the Chinese has caused anxiety among downstream countries. The Brahmaputra, however, gets most of its water from various tributaries after entering India. About 70 percent of the water volume of the Brahmaputra is generated on the Indian side of the McMahon Line through various tributaries and rainfall. The 30 percent natural flow from Tibet is still a high volume that is expected to first increase and then decline in the long-term due to the melting of the Jima Yangzong glacier, the Brahmaputra's main source.

Reports state that the Indus Water Treaty between India and Pakistan is one of the most successful water sharing agreements in the world, even surviving multiple wars between the two countries. All rivers flowing into Pakistan go through India. When Pakistan was created in 1947, it was decided that India would release sufficient water and get paid by Islamabad accordingly. The newly created country feared that India could reduce river flows, resulting in droughts downstream. With the intervention of the World Bank, the neighbors eventually sealed the treaty that gave exclusive rights to the Ravi, Beas and Sutlej to India before they entered Pakistan. In return, Pakistan received exclusive usage rights to the western rivers Jhelum, Chenab and Indus, but with some stipulations for development of projects on these rivers in India. The treaty also envisaged a commission to decide future disputes and resolve conflicts through inspections and data sharing. The treaty was put to the test to resolve the Baglihar and Kishanganga hydropower projects on the Indian side. Pakistan objected, claiming that India violated the treaty; however, arbitrators allowed the projects with some modifications. The media and opposition parties in Pakistan continue to blame India for disrupting river flows, causing floods as well water shortages in the country. Fortunately, state authorities have been mature enough to avoid such blame games, as was evident in recent incident when Pakistani media blamed India for water shortage in the country.

Reports say that Bangladesh and India share the maximum number of rivers (54) and only one comprehensive treaty was signed in 1996 that divides the Ganges' waters almost equally. The Indo-Bangladesh Joint River Commission for water management was established in 1972. However, it is the Teesta River — originating in Sikkim and flowing through West Bengal before entering Bangladesh — that is a point of contention between the two countries. In 1983, an ad-hoc water-sharing agreement was reached between India and Bangladesh, under which they received 39 percent and 36 percent of the water flow respectively. India's plan to tame the Brahmaputra through several small and big dams in Arunachal Pradesh also seems to fair well with Bangladesh. While Dhaka has raised concerns about China's plans, it does cooperate with the dams being planned in Arunachal Pradesh. Meanwhile, the opposition comes from within India, as the downstream state of Assam is concerned about the adverse impacts of these dams. It is no wonder that water has become a topic of intense confrontation among various regimes, but efforts are also being made regularly to share resources and information that diffuse tensions. As long as these mechanisms stand and are strengthened, a water war will be kept at bay.

Cross-border National Scenario: The cross-border national scenario is worse than the cross-border international scenario, more particularly in Peninsular Indian rivers where the snow based perennial flows in to the rivers is absent. They get flows during monsoon seasons and during pre and post-monsoon cyclonic activity. They are highly variable with climate change patterns.

After Independence in around two decades the regional parties dominating in State level and Central level politics. Based on who are ruling, the priorities changing and thus inter-state or inter-region sharing problems multiplying. Take for example the newly formed NDA government, it goes on changing the basic governing structure from the existing structures and spending lakhs of crores on most ineffective urban schemes such as Digital India, Swacch Bharat, etc., that benefit the party cadres at centre and state levels with least priority to irrigation projects. They have no realistic policy on either agriculture or and on irrigation wherein around 70% of the rural population primarily depending up on. They are looking at boosting urban real estate market. Thus the system is making rich, richer and poor, poorer.

4.2   Inter-state and Intra-state Disputes

Polavaram [Indirasagar] project is the classical example that comes under interstate and intrastate disputes that mar the progress of the project with huge cost escalation. Justice Bachawat Tribunal approved the interlinking of Godavari River with Krishna River through Polavaram project [it is a multipurpose project with 960 MW of power production]. With this 80 TMC of surplus water in Krishna River basin becomes allotted water. In this AP share is 45 TMC. For Polavaram Project in 80s Chief Minister Anjaih laid the foundation stone but nothing moved afterwards. In 2004, Dr. Y. S. Rajashekara Reddy [**Dr. YSR**] took over as Chief Minister of the State. He started the work [got most of the necessary approvals] including first phase of rehabilitation and right & left canals excavation before his death. Dr. YSR's government released excellent R & R Package. As part of this, Pulichintala project was initiated [I was a member of the CFE committee in APPCB which cleared the project – three collectors agreed with my proposal to give land to land for displaced]. If Dr. YSR would have been alive Polavaram project would have been completed and Rayalaseema would have benefited from the inter-linking of Godavari with Krishna. Here the water moves through the gravity from Godavari to Krishna, which is real inter-linking of rivers and has legal sanctity. Here river is inter-linked with river and not through water like in Pattiseema lift irrigation.

4.2.1   Godavari River Waters

The Godavari River rises in the Sahyadris near Triambakeswar, about 80 km from the Arabian Sea, at an elevation of 1,067 m in the Nasik district of Maharashtra state. After flowing for about 1465 km in six states in a general southeasterly direction Godavari River joins Bay of Bengal at Rajahmundry in AP. The basin extends over an area of 312813 km^2, which is nearly 10% of the total geographical area of the country. The respective drainage areas in the six states, namely Maharashtra, AP, MP, Chhattisgarh, Odessa and Karnataka are: 152199, 73201, 31821, 33434, 17752 & 4406 km^2. The number of dams constructed in Godavari basin is the highest among all the river basins in India: nearly 350 major and medium dams and barrages are constructed in the river basin by the year 2012.

In April 1969, the Central Government constituted the Godavari Water Dispute Tribunal [GWDT]. In 1980 GWDT passed an award and made Inter State Agreement on 2.4.1980 as part of the award. This is known as, Justice Bachawat Tribunal Godavari River Water distribution award. This award distributed Godavari water to riparian states after making agreements among the riparian states. Andhra Pradesh [now it is bifurcated in to Andhra Pradesh and Telangana States] got its share around 3000 TMC [allocated part is of 1480 TMC and the remaining as surplus waters]. However, so far only around 25% of this water is in use and the rest is entering the Bay of Bengal, on an average around 3000 TMC per annum. Under Jalayagnam, projects were proposed to raise the water use to around 50%. They included Pranahita-Chevella to serve drinking & irrigation needs in 7 districts of Telangana State and Polavaram, known as Indira Sagar multi-purpose project serving almost all the districts of Andhra Pradesh. In addition few other lift irrigation projects in Telangana & AP were initiated.

4.2.2 Polavaram [Indira sagar] Project

The Polavaram Project is located in Andhra Pradesh on the river Godavari, near Polavaram village about 34 km upstream of Kovvuru, Rajahmundry and 42 km upstream of (Godavari Barrage) Sir Arthur Cotton Barrage, where the river emerges out of last range of the Eastern Ghats and enters the plains [**Figure 4.1**]. The Polavaram Project is allocated water based and not surplus water based is contemplated as Multipurpose Project envisaging irrigation benefits to an extent of 7.20 Lakhs acres for the upland areas of East Godavari, Visakhapatnam Districts under Left Canal and West Godavari, Krishna Districts under Right Canal and generation of 960 MW Hydro Electric Power. In addition, this project under its left canal envisages 23.44 TMC of Water supply for industries in Visakhapatnam Township and Steel Plant, besides domestic water supply to 500 villages and towns en route and diversion of 80 TMC of water through the right canal to Krishna river to augment the supplies of Krishna Basin under inter-linking of Godavari River with Krishna River.

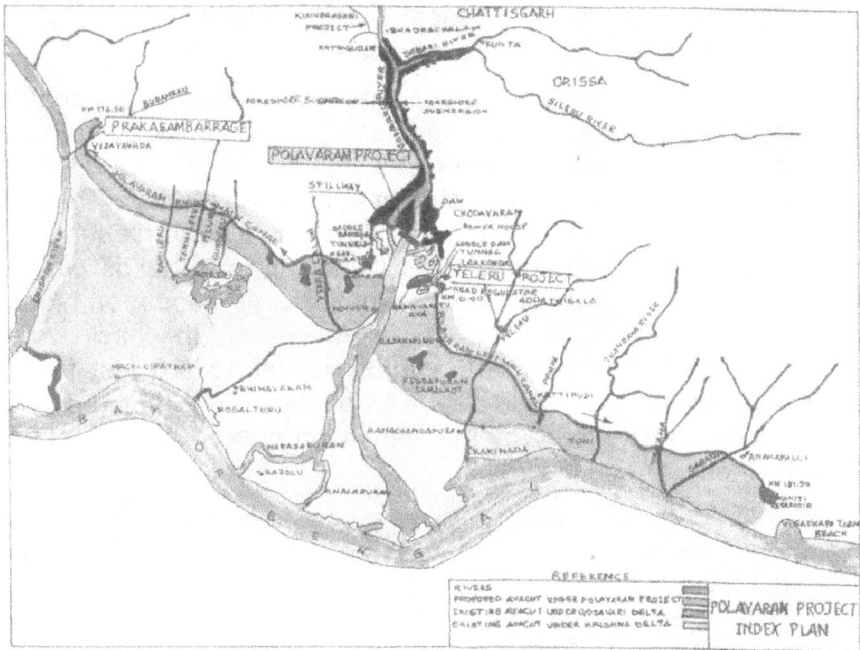

Figure 4.1: Polavaram project on Godavari river

The tribunal incorporated the agreements in their final adjudication and ordered that these agreements should be observed and carried out by all concerned. The agreement has been filed before the Tribunal and the Tribunal in clause V of their final order dated 7.7.1980 directed that the agreement be observed and carried out along with other agreements on sharing of the Godavari Waters. In accordance with the statement filed on 3.4.1980 on behalf of Government of India before the Tribunal, the Tribunal directed in Clause VI of their final order that:

1. The Polavaram Project shall be cleared by the CWC as expeditiously as possible for FRL/MWL of +150.00 feet.

2. The matter of the design of the dam and its operation schedule is left to the CWC which shall decide keeping in view all the arrangements including the agreement dated the 2nd April, 1980 as far as practicable.

3. If there is to be any change in the operation schedule as indicated in the agreement dated 2nd April, 1980 it shall be made only after consultation with the states of Andhra Pradesh, Madhya Pradesh and Orissa. The design aspects shall however be left entirely to CWC.

The State of Andhra Pradesh shall observe all safe-guards including the safe-guards mentioned in sub-clause (1) above regarding the Polavaram Project as directed by the CWC. The agreement reached with the States of Orissa and Madhya Pradesh will enable the maintenance of water levels in their territories at or below at +150.00 feet level for all floods up to 20 lakhs cusecs which are of more frequent occurrence. For floods of higher frequency ranging from 20 to 36 lakhs cusecs, the back water levels will be higher, reaching a maximum of 182.26 feet for 36 lakhs cusecs with +140.00 feet at dam site. The lands above +150.00 feet are covered by the stipulation of providing protective embankments or compensation.

In July 1941, the first conceptual proposal for the project came from the erstwhile Madras Presidency. However the project stayed idle until 2004, when the Dr. YSR-led government came to power. Now this project has been accorded national project status by the central government. This dam across the Godavari River is under construction located in West Godavari and East Godavari districts in Andhra Pradesh state and its reservoir spreads in parts of Chhattisgarh and Odessa States also. The construction activity is moving at snails speed as AP ruling TDP government is part of ruling NDA government at the centre and both don't want completion of the project. It is part of a political strategy to erase credit to Dr. YSR

Odessa, Chhattisgarh and Andhra Pradesh entered into agreement (clause vi of final order, page 80 of original GWDT) which were made part of Godavari Water Disputes Tribunal [GWDT] award. The agreement allows Andhra Pradesh to construct the Polavaram reservoir with full reservoir level (FRL) at 150 feet above the mean sea level (MSL). Odessa approached Supreme Court against the design discharge capacity of the Polavaram dam spill way stating that it should be designed for five million cusecs (cubic feet per second) which is the estimated probable maximum flood (PMF) once in 1000 years duration. Odessa argues that otherwise there would be additional submergence above 150 ft MSL in its territory during peak floods. The recorded maximum flood is 3.0 million cusecs in the year 1986 during last 115 years.

In such scenario the role of CWC is crucial to resolve the deadlock but unfortunately under the present political set up they rarely come up with unbiased answer. CWC, thus, is not doing its mandated duty and thus creating inter-state problems.

4.2.3 Cost Effectiveness of the Project

Some magazines started its tirade against the project since 2008 and continued even to date. The authors of articles titled them as "Why Polavaram is a pointless project?", "Polavaram Fraud", etc. They even made statements: the maximum flood level may even cross 90 lakh cusecs. If this is the case, whether there is dam or not, the entire area enrooted Godavari will be wiped out. Such irresponsible statements are thrown to create problems in the progress of the project. One of the article observed that the Polavaram dam involves a huge cost but its benefits will be limited. But the reality is in India the Polavaram project is the only project that is cost effective over any other project built so far. While analyzing the cost effectiveness of a project like multi-purpose Polavaram project all components must be taken in to account. Wikipedia presents a detailed report of these components to show that this is the most cost effective project.

Some people are arguing that out of 7 lakh acres of land proposed to be irrigated by Polavaram project, 4.5 lakh acres of land is already getting irrigated by some other means like Tatipudi and Pushkara lift irrigation projects. So the remaining 2.5 lakh acres of the land could be irrigated by lift irrigation by constructing alternative barrages. But by Polavaram project 7 lakh acres of land could be irrigated by gravity canals and 12 lakhs acres of additional land could be irrigated by lift irrigation (Uttarandhra sujala sravanthi, Chitalapudi and Indira sagar lift irrigation projects).The generation of 960 MW of power by Polavaram project. The other benefits of Polavaram project are diversion of 80 TMC of Godavari waters in to Krishna River to irrigate dry areas in the state, supply of 24 TMC of water to Vizag city and industries in Vizag and stabilization of Godavari delta ayacut of 10 lakh acres which is going to suffer by construction of huge lift irrigation projects in Telangana. These benefits are not possible by construction of alternative barrages.

Eng. T. Hanumantha Rao [Former Chief Engineer] proposed three barrages in place of dam. Andhra Pradesh Government sent the proposal to a technical committee to look in to the issue. The committee rejected the proposal. Member of Parliament, Palvai Govardhan Reddy from Telangana raised the issue in the Parliament. CWC evaluated the three barrages in place of a dam. CWC opposed construction of alternative barrages (instead of Polavaram dam project) to reduce submersion because by constructing barrages, only running water of the river Godavari could be diverted and there will be no storage facility of water. There will be no diversion of Godavari waters in to Krishna River, no supply of water to the industries in Vizag and no supply of water to the second crop of Godavari delta. Power generation will be reduced to 200 MW. So it is waste to invest money for construction of barrages by which only 2.5 lakh acres could be irrigated that too by lift irrigation. Now, he come up with a new proposal saying that with 30 barrages through gravity 300 TMC of water could be utilized and 4500 MU power could be produced on Godavari without any lift irrigation. Unfortunately, such people to get publicity make such statements without going in to the reality.

There is a proposal to divert 60 TMC of flood waters of the river Godavari from Polavaram project left canal near Tallapalem village in vizag district and to construct Uttarandhra Sujala Sravanthi lift irrigation project to irrigate 8 lakh acres of land in Vizag, Vizianagaram and Srikakulam districts. Indirasagar lift irrigation project is being constructed at Rudramkota village by utilizing Polavaram project back waters to irrigate 2 lakh acres of land (1.25 lakh acres in Khammam district and 75 thousand acres in west Godavari and Krishna districts). Even though the above two projects are the part of Polavaram project, they are based on flood waters of the river Godavari and they are not eligible to get National project status. So they are shown as separate projects. Original ayacut of Yeleru River was 1 lakh acres of land in E.G. district. After construction of Dowleswaram barrage in place of old anicut, 40 thousand acres of Yeleru ayacut was brought under Godavari delta ayacut by digging Pithapuram branch canal and the water saved in Yeleru ayacut was taken to the industries in Vizag. After completion of Polavaram project, the remaining 60 thousand acres of Yeleru ayacut would be brought under Polavaram project ayacut. So a new ayacut of Yeleru River could be created between Yeleru project left main canal and Polavaram project left canal to irrigate 1 lakh acres of land in East Godavari and Vizag districts.

Approximately 10 TMC of water is required to irrigate one lakh acres of land. So the water required for irrigation by Polavaram project is Irrigation by gravity canals - 70TMC; Uttarandhra sujala sravanthi lift irrigation project - 60 TMC; Chintalapudi lift-20TMC; Indirasagar lift - 20 TMC; Stabilization of Godavari delta ayacut of 10 lakh acres which is going to suffer by construction of huge lift irrigation projects in Telangana - 100TMC; Diversion of Godavari waters in to Krishna river - 80 TMC; Supply of water to Vizag city - 24 TMC. So the total amount of water required is 374 TMC. The storage capacity of Polavaram project is 195 TMC of water which is hardly sufficient for the above projects. The storage capacity of the 3 alternative barrages is below 30 TMC of water which is far below the storage capacity of Polavaram project. So Polavaram project cannot be substituted by alternative barrages.

4.2.4 Polavaram Dam Divides Politicians

The parliament has passed the AP Reorganization [Amendment] Bill on 14th July 2014, where in bill provides for the transfer of seven mandals in the newly formed Telangana state to the residuary state of Andhra Pradesh. Telangana Rashtra Samithi which formed the new Government in Telangana on 2 June 2014 has given a bandh call on 29 May 2014 protesting against the ordinance promulgated by the President.

A report in a magazine observed that "The project does not benefit the Telengana region in any way; on the other hand it will submerge villages in Khammam district". All the villages that would be submerged due to the project are under Bhadrachalam division of Khammam district. Bhadrachalam division was part of East Godavari district in Coastal Andhra until 1958. It was split from East Godavari district and added to the newly formed Khammam district in 1958 after AP was formalized. The definition of Telangana as commonly accepted includes only the Telugu speaking areas of the erstwhile Hyderabad State as on Oct 31, 1956 before the merger with Andhra State. With the bifurcation of Telangana State, the 7 mandals that submerge in Polavaram project of Bhadrachalam zone are officially merged with the Andhra Pradesh. So, submergence in Telangana region will not arise at all. In fact the Badrachalam should have been merged with AP – It is a well known fact that Sreerama Navami festival is celebrated at grandiose scale in Guntur-Tenali and not in Telangana region.

For that matter any project in India divides politicians based on their vested interests leave alone Polavaram. Politicians are always there to oppose any project proposed by ruling party but when they come back to power the issue reverses. In the case of Polavaram project, some have illegal mining interests; some others have smuggling of forest wealth interests; some have ganja cultivation interests; some others have vote bank interests; some others have tourism interests; etc. These are normal issues. Unfortunately the people who are enjoying the fruits of such illegal activities running in to thousands of crores each year and yet local people get no benefit in terms of dwelling, health care, education, etc.

In 1980 A.P., Orissa and M.P. state governments signed agreement to construct Polavaram project across the river Godavari to 150 feet height. Supreme Court appointed expert committee to see whether A.P. government violated the agreement. The committee gave a report that A.P. government did not violate the agreement because the proposed height and storage capacity of the dam are same and only the design of the dam is changed to withstand 50 lakh cusecs of flood waters instead of original 38.2 lakh cusecs of flood waters. 50 lakh cusecs of flood waters occurs very rarely say once in 500 years. So even though a case is pending in Supreme Court against Polavaram project, the court refused to give stay to stop the construction work of the project.

4.3 Submergence and Displacement

In any irrigation projects the main issue that confront the riparian states in any given river system is the submergence and displacement. NGOs, political rivals, vested interest groups, smugglers, riparian states to protect themselves from vote bank political game and monitory gain syndrome they always fight against displacement under one pretext or the other with a tag of "Tribal's & Tribal Rights".

4.3.1 Submergence

In the case of submergence, when the rivers are flowing without any obstructions such as dams, the downstream riparian states are affected by submergence with floods. When dams are built then it is reversed and thus upstream riparian states take the brunt of submergence. When downstream riparian states are affected by floods nobody bothered as it is not going to serve their vested interests.

Since the Polavaram project was initiated by Dr. YSR several national and international magazines/news services and agencies waged a war against the Polavaram project in addition to upstream states though their predecessors signed the agreement in 70-80s and the newly formed Telangana State. When Godavari experienced high floods in August 2006, submerging 370 villages for days in Andhra Pradesh, CWC made a fresh assessment of PMF. EIA says with the dam 276 villages will be submerged. That means whether there is a dam or not submergence under high floods is a common future under any river system. The later scenario is more dangerous as every time when the floods occur, houses-property is washed away with the fury of gushing waters. This is a recurring expenditure to people of those 350 villages. But, upstream states never paid any compensation to those affected people nor they said that they will control their waters. Under dam scenario to avoid frequent submergence, they will be moved permanently to a safer place with better amenities under R & R packages. In this case as long as those villages are there, it is boon to smugglers – wood & animal parts, illegal mining, ganja cultivation, tourism, etc. It is not so with the former case and thus nobody worried on that.

Thirty two years have passed after Godavari River Water Tribunal award in 1980 the upstream riparian states have not made any attempt to harness their respective allocations. The under-utilization of water is the main reason for the very high flood flows at Polavaram dam site and further in to the Sea. Without looking this aspect in 2009, CWC asked AP government modify the Dam to meet the 50 lakh cusecs based on 2006 floods of Godavari River instead of original 38.2 lakh cusecs without changing the dam height of 150 feet. This action further delayed the project progress and cost of the project has gone up.

Vast area in excess of 10,000 square km up to the Sea are frequently flooded (at least once in a decade) by Godavari floods in Andhra Pradesh. The land submergence due to Polavaram dam in Odessa and Chhattisgarh states is a fraction of Andhra Pradesh area which is affected by the floods in Godavari River. Some estimates suggest that "during the years from 1953 to 2011, Andhra Pradesh suffered nearly 558 billion rupees which is 26% of total flood damage in India". In the case of submergence of ecological sensitive zones, reservoirs generally compensate to a large extent such as groundwater recharging, cooling the environment along with increased rains. Though we lose species from the ecological zones, we get new species with reservoirs. Also in the present case government allocated funds to raise reforestation programme as stipulated by the Forests & Environmental laws.

4.3.2 Displacement

Displacement of people is not new. Nizamsagar built in 1921 displaced 13,500 people; Sriramsagar built in 1964 displaced more than one lakh people along with it; the lower Maneru displaced 13,000 people; Nagarjunasagar built in 1956 replaced 6000 people; Srisailam built in 1976 displaced 22,000 people, etc. Regarding the Tribals, under Polavaram they constitute around 50% while the same under Kurban dam is 100%, Sardar sarovar it is 57.6% & under Maneswar it is 60%.

In the case of Polavaram government of AP announced R & R packages. The then Prime Minister of India, Dr. Manmohan Singh appreciated the package and suggested to Planning Commission to evolve such packages to the whole of the Country. In fact the first batches of villages to be replaced were happy to move away even when NGOs & vested groups including political parties tried to stop them. They moved to the new rehabilitation centre. The

Polavaram project would displace 276 villages and 44,574 families spread across Andhra Pradesh state mainly, Tribal constitutes around 50% of such a displaced population. Sixty-four years after the initial conception of the project, the Government of Andhra Pradesh secured the environmental clearance from the central agency in 2005 under EIA Notification of 1994. This clearance was obtained after the state government prepared a Rs. 4,500 crore forest management plan and rehabilitation and resettlement proposal covering 59,756 hectares that were being lost under the project. In addition, Rs. 40,000 was to be allotted for each dwelling to be constructed for the displaced as against Rs. 25,000 provided by other states. Despite this clearance, the project faced political roadblocks. However, Human rights activists, NGOs, political parties, vested interests groups come out openly opposing the project on one count or the other. Meanwhile, work on the project began in April 2006 and was expected to be completed by February 2007. After 30% work of excavation work on the canals and 15% of the spillway works had been completed, the work was halted in May 2006 to seek clearance from the Ministry of Forests and Environment.

4.3.3 Disputes

Some of the disputes arise out of CWC actions in terms of raising the flood water from 38.2 to 50 lakh cusecs with 2006 floods. You can see the clearances obtained from different departments by taking in to account the Acts brought out by the Central government after EC was granted to the project. Several groups in and out commented that the project has no clearances from statutory government agencies. Let us see below on this:

- AP, Karnataka & Maharashtra agreed on transfer of 80 TMC on August 4, 1978 and the same was accepted by Justice Bachawat Tribunal;
- AP, MP & Orissa agreed for 150 ft height of the dam and associated submergence in 1978;
- Karakattas be built with 10-30 ft height over 30.2 km in Orissa and 29.12 km in Chattisghad and AP will pay Rs. 600 crores towards the cost;
- Central government in writing has given for 150 ft height of the dam – scientifically possible – on this Orissa government signed in 1980;

- MoEF clearance for Land [site clearance] was given on September 19, 2005;
- Public hearings at 5 places conducted in AP on October 10, 2005;
- EIA Report to MoEF submitted on October 19, 2005;
- EC from MoEF received on October 25, 2005;
- Forest & wildlife clearance received on July 2006;
- Central Tribal Welfare Department Rehabilitation clearance was received on August 2007;
- NEAA cancels EC on Dec 19, 2007 and AP High Court stayed this order;
- MoEF's 1st phase forest clearance given on December 2008;
- CWC Technical Advisory Committee [TAC] clearance [1st time] given on January 2009;
- Finance Commission Investment clearance [Planning Commission] was given on February 2009;
- The project getting assistance under Accelerated Irrigation Benefits Programme [AIBP] cleared on March 2009;
- Forest clearance by MoEF given on July 28, 2010 -- Forest clearance was granted with a condition that AP will construct "embarcments" to avoid submergence & displacements in Odessa & Chattisghad [CWC gave clearance for karakattas].

The neighbouring state of **Odessa** expressed its concern on the submerging of its land and decided to study this together with the officials from AP. In response Chief Minister of AP Late Dr. YSR clarified that neither Odessa nor Chhattisgarh would be affected by the construction. The problem continued until 2010, when Chief Minister of Odessa Naveen Patnaik remained steadfast in his demand for compensation and rehabilitation of tribals of his state who would be displaced due to the submerging of their land. Odessa and Chhattisgarh have filed a petition in the Supreme Court against the Project which submerges large areas of its state and allege that AP of going ahead with the project without the necessary permissions from CWC and Environment Ministry. The states also allege that public hearing in the affected areas was not held. Environmental clearance for Polavaram project was granted under EIA Notification, 1994 after AP government conducted public hearings as per this notification in

2005. AP State government agreed to pay Rs. 600 crores towards building of 'Embarcments' to avoid displacement in Odessa and Chhattisgarh.

However, Ministry of Environment and Forests [MoEF] of government of India asked AP government to conduct public hearing in those two states as per EIA Notification 2006; though it is not mandatory as per 1994 notification under which EC was granted. Also tribunal mandated embarcments and the MP & Orissa Chief Ministers signed the agreement in 80s itself. In 2009, CWC also agreed on embarcments. Whether conducting public hearings is applicable to projects cleared retrospectively is a big question. However, as directed by the MoEF, AP government wrote letters to those governments to conduct public hearing in their respective states and the cost will be borne by AP government. They did not respond on them. In such scenario, it is the duty of MoEF to conduct public hearing in those states through their regional office in Bangalore. But, they did not do that and on the contrary MoEF ordered AP government to stop work on Polavaram project. It is political game played by Jairam Ramesh the then MoEF Minister as part of AP bifurcation game where in Jairam Ramesh from Karnataka though represented Rayjasabha from AP played the pivotal role. On this, I sent mails to Jairam Ramesh and later replaced MoEF minister and Prime Minister Dr. Manmohan Singh but no response on them. Some vested groups approached the NGT and got stay but this was lifted by AP High Court. Now, the NDA government lifted the ban imposed by MoEF. Here politics played over law.

It is argued on the submergence in Odessa & Chhattisgarh the "Odessa and Chhattisgarh entered into agreement (clause 3e, Annexure F, Page 159 of original GWDT) to construct a Hydro electricity project at Konta / Motu just upstream of the confluence point of Sileru tributary with Sabari River (tri-junction point of Andhra Pradesh, Odessa and Chhattisgarh borders). When this project is constructed, the land submergence would be more than that of Polavaram back waters. It would be better that Odessa and Chhattisgarh enter into agreement with Andhra Pradesh to shift the location of this Hydro electricity project further down stream in Andhra Pradesh territory to harness Sileru River water also for hydro electricity generation. This joint project of the three states would eliminate the back waters issue of Polavaram dam."

The areas of submersion by Polavaram project in Odessa and Chhattisgarh states are minimal and A.P. state government is ready to construct embankments along the banks of Sabari and Sileru Rivers and arrange pump sets by the side of the embankments to drain excess water in rainy season so that not even a single acre of land would be submerged in Odessa and Chhattisgarh states by Polavaram project. Otherwise A.P. government wants to pay compensation for the submerged lands in those states. Choice is left to those states. After construction of Polavaram project old lift irrigation projects like Tatipudi and Pushkara lift irrigation projects could be removed to relieve burden on Doweleswaram barrage & power utility and the new lift irrigation projects like Uttarandhra sujala sravanthi and Indirasagar lift irrigation projects which are based on Polavaram project could be continued.

4.4   Practical Issues and Problems

4.4.1 Progress in Projects

The progress of projects primarily depends up on the builder. Here there are basic issue that decide the progress of the project, namely (1) illegally selling the basic material such as steel, cement, etc., that were provided at controlled price to the builder by the government, (2) through delays manipulated cost escalation, and (3) with poor technical knowledge. If the government and bureaucrats get their share, the builder further delays. This is exactly what is happening with Polavaram project. A TV channel presented how the present builder is dodging the construction as he is ruling party MP. He appears following the dictates of ruling junta and delaying the construction activities. Now AP government agreed to pay the present cost and divert some work to German Company. No government agency took this issue!!!

4.4.2 Implementation of Rehabilitation and Resettlement [R & R]

In the rehabilitation and resettlement issue there are several practical issues as well problems. Some are government implementation and some others are associated with vested group games. We have seen the reports saying that the rehabilitation and resettlement was still not completed in the case of Narmada project and Srisailam project ousters. In the case of Jurala project the compensation was got through court twice. In such scenario, government must take pragmatic decision without pending the cases as otherwise the future projects will be severely hampered. However, delays in the execution of irrigation projects, the cost goes up sky rocketing like in the case of Narmada and now with Polavaram. The latest estimates suggest the cost may go from Rs. 14,000 crores to Rs. 30,500 crores in which the major share goes towards land acquisition and R & R.

Re-forestation Programme: CAG reported that the reforestation programmes undertaken by the Forestry department of MoEF is highly misused. It says the forest officials created fictitious NGOs on their family members and the contract were shown on those NGOs name. The amount is misused this way without really developing reforestation programmes. This is like fence eating the farm. If MoEF has earnestly implemented the reforestation programmes, India would have increased area under forest cover significantly.

Displaced Families: In the case of counting of displaced families the main problem is associated with the NGOs and Political parties. They bring in new families. The government refuses to consider them for compensation or providing them resettlement packages. This tussle continues through court cases in many a times.

While preparing the project they calculate the number of families to be rehabilitated and compensation issue will be decided on that as per R & R package. But, at the time of real rehabilitation, the number grows in terms of division of families, people later settled with the help of NGOs and political parties or shifted from Naxalite problems or deforestation problems, or mining problems, etc. Even with the new Land Bill of 2013, the same will be there. Unfortunately here governments have not formulated unambiguous system of census both in terms of humans and land.

4.5 Concluding Remarks

With regional parties taking centre stage at states and central government levels, the projects with interstate or intrastate took decades to settle the even minor disputes. These will hamper the benefits to accrue to people but politicians are least bother on such issues but their primary goal is get power – vote bank – and monetary gains. To avoid future water wars, the government must fix the loopholes in all the activities related to irrigation and irrigation projects. Without that there will not be speedy progress in irrigation and irrigation projects in India.

The Polavaram project estimates in 2005 was Rs. 10, 271 crores with maximum flood water of 38.2 lakh cusecs. The CWC in 2009 raised the maximum flood water to 50,000 lakh cusecs. With this the revised estimates of the project in 2011 was Rs. 16,000 crores. On 11th October 2015, AP Cabinet decided to raise the cost to Rs. 36,000 crores. All this is happening because the central government has no clear cut planning.

Future of Water Resources in India

5.1 Introduction

Under the deteriorating standards in politics, we are encountering severe problems in accomplishing the goals in irrigation sector at state and central levels. NDA & UPA, for that matter any other political systems in India, are corrupt to the core. With this scenario the major causality is water resources utilization – irrigation projects. Though it is prevalent all over India, let me present a case as an example from a report which states that the newly formed Telangana State decided to cancel all those contracts relating to 96 irrigation projects initiated by the previous governments and call for fresh bids. With this, the completion of projects will be delayed further; and thus becomes a suicidal plan. Also, Telangana Government in a hurry to go ahead with "Palamuru - Ranga Reddy" lift irrigation scheme [expected to lift 90 TMC in 90 flood days – where from this water comes is a big question] at a cost of around Rs. 35,200 crores. At the same time not putting thrust on completing the projects like Kalwakurthy, Bhima, Nettempadu & Koilsagar in the same area, which have completed 80 to 85% of the work and expected to provide water to 7 lakh acres. Under this, "Might is Right". NDA government at the Centre hasn't made any attempt to stop such actions on such major issues even though it is state's subject, nor the CAG bothered to look in to this fiasco running into few lakh of crores.

However, keeping the growing population and thus increasing water needs in mind, as a first step the governments must reduce the water going in to the Oceans and the Seas as waste through building dams and interlinking of rivers wherever possible with minimal damage to environment/ecology. This will benefit pollution free agriculture systems to meet the food needs. However, there is strong opposition to such schemes, from different quarters. Some are arguing that "Why should the entire rainfall come in to a dam and then be diverted to other areas? Catching rain where it falls should be the mantra". Our traditional farm bund system comes under catching rain where it falls. This system allows under rainfed condition to improve soil moisture in the root zone.

In the 7th and 8th Five-year Plans, government spent thousands of crores on watershed programmes. This was found a wasteful expenditure as it works under stable rainfall conditions like in MP and failed in dry areas of AP, Gujarat, Karnataka and Maharashtra. However Anna Hazare and Waterman tried to show they are successful without understanding the rainfall variations in space and time in India that occur over a limited period. To get successful agriculture and water resources availability we need storage facilities – even from few centuries this system existed in India, see Mahanandi Temple & Srisailam temple in Kurnool District and even Tirumala the ebode of Lord Venkateswara. Here the most important component is the climate change – not the global warming. Before we make statements, we must understand the system. India needs dams and interlinking of rivers to meet the growing needs of ever increasing population and urbanization. Good and bad are like a coin has head and tail; but if it provides succor to majority, we must give weight to good and try to minimize bad. We must follow this in water use.

5.2 Inter-linking of Rivers in India

5.2.1 Introduction – Benefits and Problems

Interlinking of Rivers program [**ILR**] aims to transfer water from surplus to water deficit areas in the country. It will help saving the people living in drought-prone zones from hunger and people living in flood-prone areas from the destruction caused by floods. The idea of ILR has its origin a century and half back. During the British raj, an engineer **Sir Arthur Cotton** had sought to link the Ganga and the Cauvery to improve connectivity for navigation purposes but due to the increased railway connectivity among the areas, the idea was shelved.

Transferring water from surplus Himalayan Rivers to water-scarce regions in western and peninsular India linking of Ganga and Cauvery below Patna in 1972 was proposed by Union water resources minister K. L. Rao. In 1982, Prime Minister Indira Gandhi set up the National Water Development Agency [NWDA] to take the plan further. NWDA is an autonomous body to carry out the water balance and feasibility studies of the river linking program.

The first NDA government (1999-2004) was keen to implement the ILR, and the Supreme Court, following public interest litigation, in 2003, asked for it to be implemented by 2016. During UPA [2004 to 2014] ILR hasn't moved due to interstate disputes and opposition from environmental groups. In February 2012, Supreme Court, gave its go-ahead to the ILR and asked the government to ensure that the project is implemented expeditiously. Due to reluctance of certain states, the Centre has not been allowed to undertake detailed surveys. For example Tamil Nadu favoured ILR as no major river originates in the state and it is dependent on interstate rivers. Assam, Sikkim and Kerala opposed ILR as they want exclusive right to use their water resources and that such transfer should not affect any rights of these states. The present NDA government has stated its intentions to go ahead with the project and has formed a special committee for the same.

Reports state that the expected benefits from ILR include irrigating 35 Mha, enabling full use of existing irrigation projects and generating power to the tune of 34,000 mw with added benefits, including flood control. The cost of the project was estimated at 5,60,000 crore. This exercise, with a combined network of 30 river-links and a total length of 14,900 km (9,300 mi) at an estimated cost of US$120 billion (in 1999), would be the largest ever infrastructure project in the world -- ; the true cost can be known only when the detailed project reports of the 30 river link projects are drawn up and as well commencement of such projects in real time.

Some reports observed that to better utilize the water that is entering the Sea as waste, there are proposals for the implementation of the inter-basin water transfer link schemes that are taken up in a phased manner depending on the priorities of the government. The links namely: (i) Ken-Betwa link, (ii) Parbati-Kalisindh-Chembal link, (iii) Godavar (Polavaram) – Krishna (Vijayawada) link, (iv) Damanganga-pinjal link and (v) Par-Tapi-Narmada link have been identified as priority links for consensus building amongst concerned States for taking up preparation of detailed project reports. **Figures 5.1[a, b & c]** present the schematic representation of interlinking of rivers at [a] Peninsular, [b] Himalayan and [c] national sectors.

PROPOSED INTER BASIN WATER TRANSFER LINKS
PENINSULAR COMPONENT

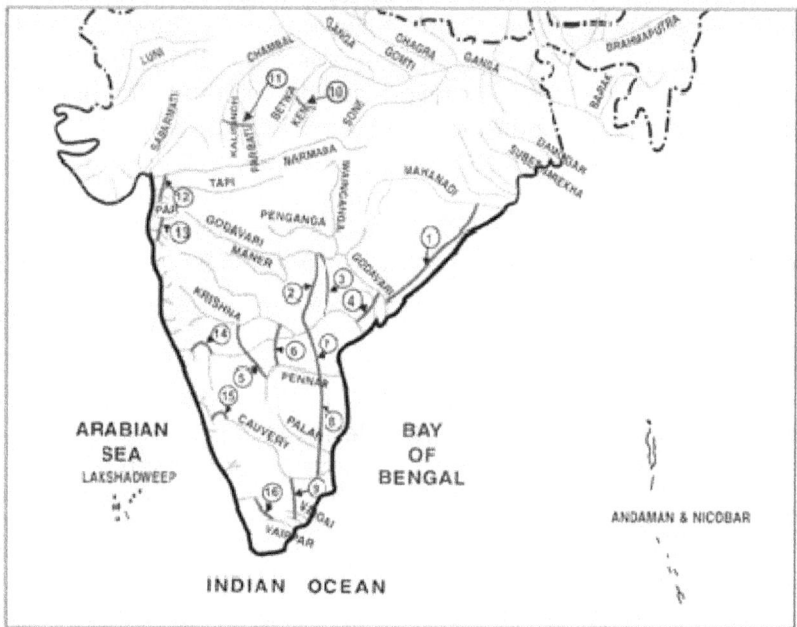

1. Mahanadi (Manibhadra) – Godavari (Dowlaiswaram) *
2. Godavari (Inchampalli) – Krishna (Nagarjunasagar) *
3. Godavari (Inchampalli) – Krishna (Pulichintala) *
4. Godavari (Polavaram) – Krishna (Vijayawada) *
5. Krishna (Almatti) – Pennar *
6. Krishna (Srisailam) – Pennar *
7. Krishna (Nagarjunasagar) – Pennar (Somasila) *
8. Pennar (Somasila)–Palar– Cauvery (Grand Anicut) *
9. Cauvery (Kattalai) – Vaigai – Gundar *
10. Ken – Betwa *
11. Parbati – Kalisindh – Chambal *
12. Par – Tapi – Narmada *
13. Damanganga – Pinjal *
14. Bedti – Varda
15. Netravati – Hemavati
16. Pamba – Achankovil – Vaippar *
* FR Completed

(a)

PROPOSED INTER BASIN WATER TRANSFER LINKS
HIMALAYAN COMPONENT

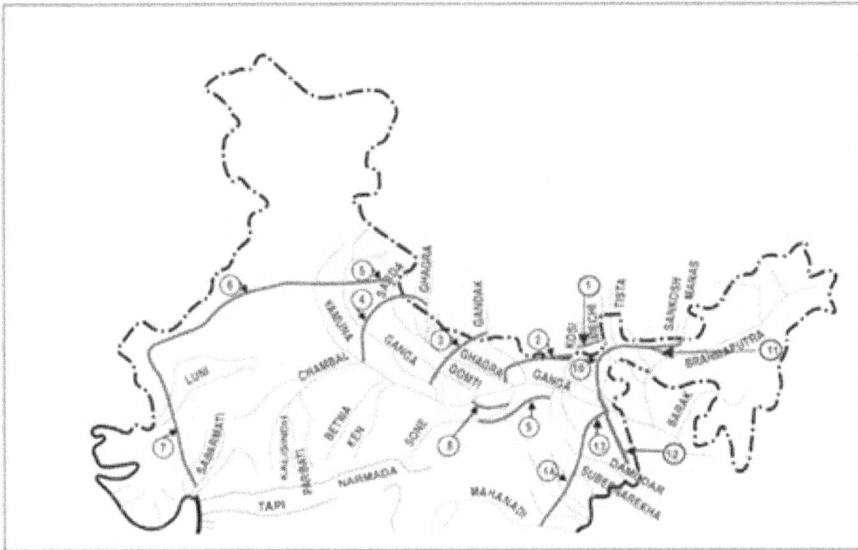

1. Kosi – Mechi	8. Chunar- Sone Barrage
2. Kosi – Ghagra	9. Sone Dam – Southern Tributaries of Ganga
3. Gandak – Ganga	10.Manas –Sankosh - Tista - Ganga
4. Ghagra – Yamuna *	11.Jogighopa – Tista – Farakka (Alternate)
5. Sarda – Yamuna *	12.Farakka – Sunderbans
6. Yamuna – Rajasthan	13.Ganga (Farakka) – Damodar – Subernarekha
7. Rajasthan – Sabarmati	14.Subernarekha – Mahanadi
	* FR Completed

(b)

The Grand Interlinking Blueprint

Penisular Component
1. Mahanadi - Godavari
2. Inchampalli - Nagarjunasagar
3. Inchampalli - Pulichintala
4. Polavaram - Vijayvada
5. Almatti - Pennar
6. Srisailam - Pennar
7. Nagarjunasagar - Somasila
8. Somasila - Grand Anicut
9. Kattalai - Vaigai - Gundar
10. Ken - Betwa
11. Parbati - Kalisindh - Chambal
12. Par - Tapi - Narmada
13. Damanganga - Pinjal
14. Bedti - Varda
15. Netravati - Hemavati
16. Pamba - Achankovil - Vaippar

Himalyan Component
1. Kosi - Mechi
2. Kosi - Ghagra
3. Gandak - Ganga
4. Ghagra - Yamuna
5. Sarda - Yamuna
6. Yamuna - Rajasthan
7. Rajasthan - Sabarmati
8. Chunar - Sone Barrage
9. Sone Dam -
Souther Tributaries of Ganga
10. Manas - Sankosh - Tista - Ganga
11. Jogighopa - Tista - Farakka(Alternate)
12. Farakka - Sunderbans
13. Ganga (Farakka) - Damodar -
Subernarekha
14. Subernarekha - Mahanadi

Source: IWMI

(c)

Figure 5.1: Proposed ILR systems: [a] Peninsular component, [b] Himalayan component and [c] All-India component

Detailed project reports for five links, including Ken-Betwa and Par-Tapi-Narmada, have been prepared. So far only Ken-Betwa project is under survey. In 2005, MoU was signed between Union Water Ministry and, CMs of MP & UP. Approximately 8,650 ha of forest land in Madhya Pradesh are likely to be submerged for the project; and part of that forestland is a part of the Panna National Park. Here we must remember one important fact – to create settlements to serve political interests, the political parties are butchering forest areas in thousands of acres. We rarely account this.

5.2.2 Centre – State Legality Issues

Subject "water" is placed in the Constitution in Entry 17 of List II (State List) of Schedule VII. However, the caveat is Entry 56 of List I (Union List), which says, "Regulations and development of interstate rivers and river valleys to the extent to which such regulation and development under the control of the Union is declared by Parliament by law to be expedient in the public interest." Article 246 and 7th Sch. the states have exclusive jurisdiction over waters in their territories, including interstate rivers. This provision prevents Union and Judiciary to settle the issue. It has also stopped the Centre from clearly defined water rights among states, and ends the long drawn legal battles. The latest example is the second Krishna Water Disputes Tribunal. It has turned into a warzone, with a battery of lawyers, technical staff and irrigation department officials from Maharashtra, Karnataka and Andhra Pradesh. Everyone is fighting to win the maximum allocations of the Krishna River for their respective states. However, in this case the main concern is it not pertains to legal issue but it is a technical use in which the tribunal fraudulently favoured Karnataka at the cost of Andhra Pradesh. Unfortunately the court also is not looking in this direction even though I submitted the details on the "Technical Fraud" enacted by the tribunal. In such issues the Irrigation Ministry and CWC can intervene to resolve the technical problems but they are least interested – vote bank politics.

Convention on the Law of the Non-Navigational Uses of International Watercourses [CLNNUIW]: it is a document adopted by the UN on May 21, 1997, pertaining to the use and conservation of all waters that cross international boundaries, including surface and groundwater. Unfortunately, the convention is not yet ratified. India, US, China, Canada and Australia are major opponents of the CLNNUIW.

China has several projects in west-central Tibet that may reduce the river water flow into India + Bangladesh. China is planning to divert 0.200 bcm of the Brahmaputra from south to north to feed the Yellow River. If this is true, India will face a severe crisis, many of the hydel projects in the Northeast India will have to be stopped. 0.600 out of the 1.900 bcm of river runoff in India comes from Brahmaputra. You can imagine what would happen if the bulk of this is diverted by China.

It says: India is faced with poor water supply services, farmers and urban dwellers, all pumping out groundwater through tube-wells. It has led to rapidly declining water tables and critically depleted aquifers, and is no longer sustainable. Government gives highly subsidized or even free electricity. Hence farmers are using pump sets in lax manner. India is getting seriously water-stressed; and we need to act fast. Water has to be treated not as a local resource, but a global resource. We need to see if a change in its constitutional status is required. We also need to enhance our water-storage capacity, as we suffer the most from the vagaries of the monsoon. River-linking project, alongside a chain of water-conservation projects, would offer a solution.

5.2.3 Reports on Inter-linking of Rivers

The Times of India [6th October 2015] presented a report "Centre tasks AP to make DPR for linking of rivers: Mega project is touted as the biggest in India". This report observed that the central government enthused over the Pattiseema and the commissioner of Union Water Resources Ministry, in a letter, asked the AP irrigation authorities to prepare a detailed project report (DPR) for Krishna - Pennar, Krishna (Alamatti) - Pennar and Pennar - Cauvery (Grand Anicut) linking projects [**Figure 5.2**]. The projects involve Andhra Pradesh, Tamil Nadu, Maharashtra, Pondicherry and Karnataka states.

A national perspective plan formulated by the Centre envisages diversion of surplus flows of the Mahanadi basin and the Godavari basin to the water scarce Krishna, Pennar, Cauvery, Vaigai and Gundar basins in the South. The report also observed that the NWDA has assessed the water balance position in various peninsular river basins keeping in view the ultimate development scenario in these basins. Based on these studies, NWDA has formulated proposals for diversion of 430 tmc ft of water annually from Mahanadi through Mahanadi-Godavari link canal. From Godavari, a quantity of 925 tmc ft of water (including water brought from Mahanadi) is envisaged to be diverted to Krishna River through three links via Inchampalli-Nagarjunasagar, Inchampalli-Pulichintala and Polavaram-Vijayawada. Out of this water brought from Godavari, a quantity of 498 tmc ft is envisaged to be diverted from Krishna to Pennar through three link canals via Almatti-Pennar, Srisailam-Pennar and Nagarjunasagar-Somasila. From Pennar, a quantity of 303 tmc ft of water is proposed to be diverted

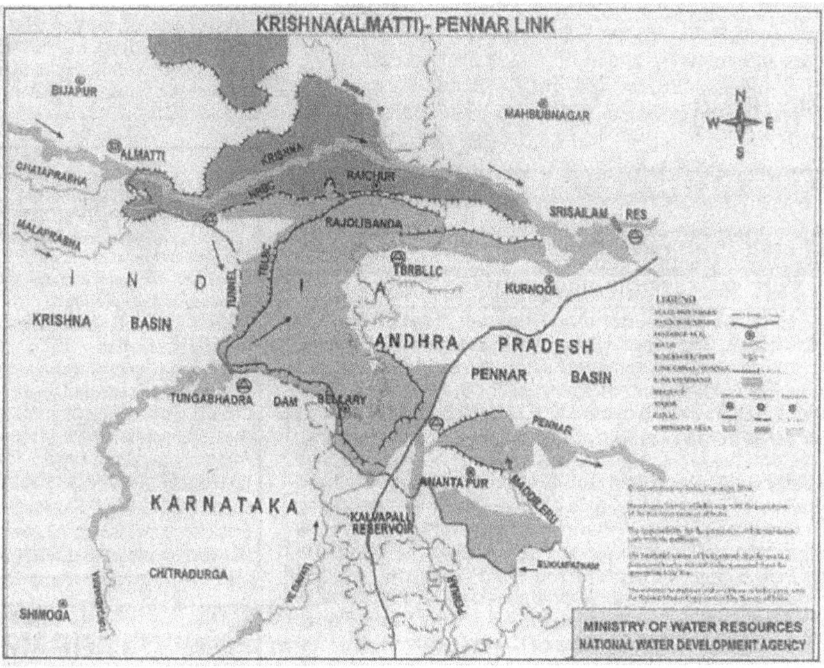

Figure 5.2: ILR Southern peninsular – AP link

to Cauvery River through Somasila - Grand Anicut link and further down south Cauvery - Vaigai - Gundar link canal to meet the demands of areas lying below Cauvery River up to Gundar basin in Tamil Nadu State. The report also observed that "Andhra Pradesh was given major responsibility of preparing the detailed project report of three major river linking schemes that would benefit AP and Tamil Nadu. The Pennar (Somasila) -Palar - Cauvery link project envisages diversion of 303 tmc ft of water from the existing Somasila dam across Pennar river in Nellore district by transporting water through 529.190 km stretch," as told by water resources department principal secretary Adithyanath Das.

The canal will run parallel to the existing Kandaleru flood flow canal for about 10 km and through Kandaleru - Poondi canal upto 80 km and then traverses a distance of 439.190 km before joining Grand Anicut across Cauvery in Thanjavur district of Tamil Nadu. The link canal traverses through Nellore, Chittoor districts of Andhra Pradesh; Tiruvallur, Kancheepuram, Vellore, Tiruvannamalai, Villupuram, Cuddalore, Perambalur and Tiruchchirappalli districts of Tamil Nadu, passing through the river basins of Pennar, streams between Pennar and Palar, Streams between Palar and Cauvery to reach Grand Anicut. The existing Somasila dam is proposed to be utilized as off take of canal for the planned diversion. The project cost was estimated at Rs 6,769 crore. At the same time, AP government will prepare diversion of Krishna water from Almatti to Pennar to give water to drought-prone areas of Anantapur district. The project cost is pegged at Rs 6,600 crore and expected to provide irrigation to 2.5 lakh hectares. The report further observed that "AP has shown the country the right model to execute river-linking projects. We have completed the Pattiseema project in a record 158 days and also started releasing water through Polavaram right main canal," as told by the AP water resources minister D Umamaheswar Rao. The third project is the Krishna (Srisailam) - Pennar River linking which has already been executed through Telugu Ganga project except for building four mini hydel power projects. The hydel projects are proposed en route of the conveyance system utilizing the natural falls of the streams. Will this materialize by giving responsibility to one state instead central government doing the same? We have seen the data scenario fiascos in **Chapter 3**. Unless these issues are resolved, it becomes a futile and becomes a war zone. Also, AP is unable to resolve a minor issue on Polavaram with Odessa and Chhattisgarh, how could this big issue connecting several states will be solved?

Gopal Krishna an expert working on ILR presented an article titled "Inter Linking of Rivers: How to kill rivers, the Gujarat way" in "ecoearthcare" magazine in April 2014. He argued that "BJP's Prime Ministerial candidate is arguing that 'In Gujarat, we have interlinked 20 rivers and the agriculture growth is 10 percent.' This claim seems to be misleading because government's statistics reveals that agriculture in Gujarat is largely dependent on groundwater for irrigation." I remembered a TV discussion on national channel on 'Watershed projects'. The member from NBA argued that in Gujarat agriculture development took place with watershed projects. This was countered by the irrigation minister of Gujarat state. He argued for Narmada project for Gujarat agriculture development. NBA, Anna Hazare, Waterman, etc., argued for watersheds over dams. In fact I countered on the 'watershed programmes' as with this programme politicians-bureaucrats nexus pocketed thousands of crores from this programme. The watershed programme was initiated in the 7th Five-year plan with a limit of 750 mm average rainfall but in 8th plan this was lifted. This helped them to amass wealth. At a felicitation meet to Journalist from Mumbai, SaiNath at Sundarayya Vignana Kendram in Hyderabad, senior editors of daily news papers made an observation that with watershed projects, people who used to move by bicycle are now moving on BMW cars. Waterman argues for watershed based on MP experience with stable rainfall to other zones such as AP, Maharashtra, and Karnataka & Gujarat with unstable rains it failed. NBA dumped all dam projects in to a basket and opposed and in the same manner Gopala Krishna dumped all ILR projects in to a basket and put forth arguments against ILR system. He argued that the interlinking of rivers is a misnomer – "It is actually an exercise aimed at diversion of rivers from their centuries old courses". But this is not so in real terms, *but only a part of its water is diverted to other rivers and the main river continues its existing journey.* He further made several such statements but majority of them are inconsistent with what he says in different paragraphs and generalized statements will not apply to all IRL projects.

He argued that 'The Himalayan rivers linking data is not freely available, but on the basis of public information, it appears that the Himalayan rivers linking component is not feasible for the period of review up to 2050.' The report underlines that the problems are in the entire plan of linking the Himalayan Rivers" *By putting the data in public domain may resolve this issue but however, the problem here is that such action may create negative impacts from neighbouring countries – we must take note of practical issues.*

He further argued that in order to comprehend the claims of rivers being "surplus" take the case of Ganga which is deemed as a "surplus" trans-boundary river from which water is planned to be removed to relieve flood by means of barrage-canal works for transfer to Subarnarekha-Mahanadi-Godavari-Krishna-Pennar-Cauvery. The latter rivers' flow during monsoon flood is at the average rate of 50,000 cumecs. This will create an ever present disaster. If the flood is to be relieved, water in substantial quantity needs to be removed by means of the link canals that will "be 50 to 100 m wide and more than 6 m deep", according to government's website explaining the modus operandi of "benefits." When a 10 m deep 100 m wide lined canal can at most carry about 1,500 cumecs of water, that would relieve flood only to the extent of 3 per cent and that too only downstream of the canal. This is the landscape in which water from so called "surplus" rivers is to be transferred to so-called deficit rivers – we must remember that *practical issues will come up only when individual projects are taken up.* He argued that 'You claim that agriculture growth rate in Gujarat is 11%, but by your own government's estimates in 2006-2007 agricultural production in the state was Rs. 27,815 crore. In 2012-2013, agricultural production fell to Rs. 25,908 crore. This means agricultural production has fallen in Gujarat during your tenure and the annual agricultural growth rate is −1.18%. How do you then claim agriculture growth rate is 11%?'; and 'The height of Narmada Dam was raised in 2005 to provide water to the people of Kutch for drinking and farming. But, even eight years later, the people of Kutch have not got water. This water was given to some of your favourite industrialists. Why this discrimination against the people of Kutch?'

However, with reference to issues raised by Gopal Krishna on Gujarat ILR projects, some are important issues that need an answer from Gujarat Government and from the present Prime Minister, Narendra Modi. Though Gopala Krishna presented excellent report on interlinking of Indian Rivers but what we need now is give priority to those interlinking projects with minimal disputes and problems. We have to start somewhere other-wise politicians wasting public money to serve their vested interests on some short lived lift irrigation and shout from the roof-top 'I accomplished interlinking of rivers. This must be stopped'.

5.2.4 Practical Issues

The Economic Times [September 15, 2015] reports that: "The Centre today announced that the Ken-Betwa link would be a model for its ambitious river interlinking project and also that steps are being taken to link Sankosh and Mahanadi. A statement here said that Union Water Resources Minister Uma Bharti, while addressing a gathering on the occasion of the sixth meeting of Special Committee for inter-linking of rivers here, said that various clearances for Ken-Betwa link project are in the advance stage of processing. The government will start implementing this national project as model link project of Interlinking of Rivers programme after obtaining statutory clearances,' she said and stressed the need to inter-link rivers for providing water to water-short, drought-prone and rainfed areas. She said that the government is committed to implementing the programme with the consensus and cooperation of concerned state governments. Stating that the Detailed Project Report of Par-Tapi-Narmada link project has been completed by National Water Development Agency (NWDA) and submitted to governments of Gujarat and Maharashtra on August 25, she said that this is the third link project after Ken-Betwa and Damanganga-Pinjal for which the DPR has been completed. Bharti also said the issue of water sharing between Gujarat and Maharashtra in respect of Damanganga-Pinjal and Par-Tapi-Narmada link project is now required to be taken up on priority. 'I would therefore urge both the governments of Gujarat and Maharashtra to address the issue of water sharing and arrive at an agreement so that the implementation of these two projects can be taken up at the earliest,' the statement quoting the Minister said. Bharti said that task force for interlinking of rivers constituted by the Ministry has started its working and this will help in bringing speedy consensus among states on the link projects. She also said that a team of senior officials of her Ministry led by Special Secretary had a meeting with Chief Secretary of West Bengal at Kolkata recently and discussed the proposal of Sankosh-Mahanadi link system which comprises four links namely Sankosh-Teesta-Ganga, Ganga-Damodar-Subarnarekha, Subarnarekha-Mahanadi and Farakka-Sunderbans. The proposed link system would provide large irrigation benefits of about 10.5 Lha besides domestic/industrial water supply to West Bengal. The state government has been requested to agree to the proposal and furnish their suggestions for its improvement, the statement said. Speaking at the occasion, Karnataka Water Resources Minister M B Patil said that concerns of his state over inter-linking of Mahanadi and Godavari rivers have not been taken into consideration. According to a statement, referring to the restoration of the state's share of water from surplus water of Mahanadi and Godavari, Patil said despite persistent efforts of Karnataka, the concerns of his state have not been brought into focus by either the Ministry of Water Resources or River Development and Ganga Rejuvenation or NWDA. He said that Pre-Feasibility Reports (PFR) of Ponnaiyar (Krishnagiri)-Palar link in Tamil Nadu as prepared by NWDA has taken into account an import of 271 Mcum (9.57tmc) up to the upstream of Krishnagiri dam available from drinking water supply of Bangalore city. In view of the right to use the regenerated water by Karnataka from Bangalore water supply which is being drawn from Cauvery River, it would have been appropriate not to have considered the 271 Mcum in the water balance study in favour of Tamil Nadu, the statement added. Patil was of the opinion that reckoning of the regeneration flow by NWDA in its water balance study ignoring Karnataka's interests is not correct, the statement said. The statement also said that Minister of Water Resources of Jharkhand Chandra Prakash Chaudhary stated that the PFR of Sankh-Southkoel and Southkoel-Subarnarekha river link projects prepared by NWDA has been put on hold because of some objections by the Government of Odessa. Minister of Water Resources of Maharashtra Vijay Shivtare said that with reference to Damanganga-Pinjal link project his state should be allocated water at 75 percent dependability instead of 90 per cent dependability and as such, the size of the tunnels/water conveyance system should be designed to suit diversion of such quantum of water, ensuring optimum use of diverted water. *He also suggested that during water deficit year, distress should be shared by Maharashtra and Gujarat in proportion of water allocation – the Justice Brijesh Kumar Tribunal did not follow this while distributing water to Maharashtra, Karnataka & Andhra Pradesh.* Senior officials from various states including West Bengal, Haryana, Punjab, Tamil Nadu, Gujarat, Madhya Pradesh, Uttar Pradesh, Andhra Pradesh, Odessa, Chhattisgarh and Assam also attended the meeting, it added."

NGO groups argue that "If Ken Betwa link is the model of ILR then ILR has no future considering the number of violations, severity of impacts, flawed impact assessment, public hearing protests curbed, manipulated water balanced, neglect of upstream, destruction of Tiger Reserve and now flawed appraisal." If the inter-linking of rivers expected to move without much problem, the government must address these deficiencies.

NWDA/CWC must analyze such scenarios in the light of climate change and other monsoon related aspects and come up unbiased and unambiguous solutions to minimize interstate and intrastate disputes to a maximum extent. Otherwise the disputes will limit the progress of ILR.

5.3 Godavari-Krishna Rivers – Linking Projects: A Case

General issues: Under ILR, the Godavari River basin is considered as a surplus one, while the Krishna River basin is considered to be a deficit one. As of 2008, 644 TMC of underutilized water from Godavari River flowed into the Bay of Bengal. Based on the estimated water requirements in 2025, the Study recommended that sizeable surplus water was to be transferred from the Godavari River basin to the Krishna River basin. Government decided to re-look at Godavari-Krishna Rivers IRL in terms water availability. NWDA decided to get the latest water availability figures though it was done in 2005. They expected this will be over by November 2015. They are now indirectly questioning even the Justice Bachawat Tribunal award. The hidden agenda in this appears to be to provide water immediately to capital city building and to achieve this goal, "Polavaram project must be slow downed at the behest of Ruling party in AP". They need water for the new capital & associated real estate building and this is achieved through Pattiseema lift irrigation as farmers may not like diverting Krishna water or Polavaram water for these activities. However, TDP government goes on trumpeting that this water is for Rayalaseema.

However, it appears that in all these assessments they haven't looked in to the climate change [see **Chapter 3**] related flows and cyclonic activity related flows. Unless they are taken in to account, we may not achieve sustainable flows. Sir Arthur Cotton constructed a dam across the river Godavari near Dhavaleswaram in 1847-52 and on the river Krishna at Vijayawada in 1852-55. They have limited storage capacities. To protect these two barrages from floods fury, the levels were maintained. The flood water beyond the levels, were released in to the Bay of Bengal, which becomes a wastewater.

For interlinking of rivers we need good storage facility. This is not so in the case of Dhavaleswaram barrage. Pattiseema has to use water from the limited water storage at Dhavaleswram barrage after providing water for the existing ayacut and drinking/industry needs. The limit is put at 15 m and thus only when the water level is above this limit, Pattiseema can pump the water. Also, to protect the barrage when water level crosses the fixed limit, the flood water is released in to Bay of Bengal. This is very important during cyclonic activity time. In fact, in 2008, though received good flows to Dhavaleswaram barrage, by the time Rabi crop was sown there was no water to protect the Rabi crop in the Godavari dealt area. Dr. YSR through an innovative idea "by digging bore wells in the river Godavari bed and pumped the water" supplied water to save the Rabi crop. This has given bumper crop production. With the Pattiseema the number of flood days are fixed and based on that 80 TMC water transfer was decided. This is a big question; will this be achieved under such a scenario?

However, this is not the case with Polavaram. It is a substantially high water holding reservoir and that too is built on allocated water and not on surplus water. This helps the flow through gravity through Polavaram right canal that serve the interlinking of Godavari River with Krishna River. AP share in the 80 TMC is 45 TMC only and the rest goes to other states. This will be provided to Rayalaseema through Pothireddypadu head regulator at Srisailam dam when the water reaches beyond 854 ft – full capacity is 885 ft and dead storage level is 834 ft. Here unfortunately both AP and Telangana governments go on producing power even below dead storage level, how the 854 ft level of water is achieved in bad rainfall/flood flow years? And then how the water flows into Rayalaseema, is a big question? Also, at present Pothireddypadu is not ready to carry water; and the projects in Rayalaseema and south coastal Andhra to use that water are not ready. Unless these are put in to a separate body to regulate, it is rarely possible that Rayalaseema get its share from Polavaram leave aside the Pattiseema. Here it is pertinent to say that the surplus water in Krishna River as per Justice Bachawat award were illegally used beyond Srisailam dam instead building the 8 projects proposed for using the surplus water. The big question is will those who are using surplus water illegally for the last more than 25 years on an average around 276 TMC will allow Rayalaseema & south Coastal Andhra Pradesh to get it share with the new capital coming up in Guntur-Vijayawada???

Pattiseema Lift Irrigation: ILR by definition, the water from One River joins another river through gravity. Lift irrigation does not come under ILR. It will not link rivers but only link water. When rivers are linked, along with the water the life forms in one river move in to another river. So, water taken from Godavari through Tadipudi or Pattiseema through lift irrigation will not connect the two rivers through water flows. Unfortunately no irrigation expert says this is not an ILR. This is our political experts!!! Lift irrigation projects don't meet this theory. In the lift irrigation, nowhere the rivers are really/physically connected. A lift irrigation scheme was planed during Dr. YSR time to bring 170 MGD of Godavari water to Hyderabad to meet its drinking water needs. The first phase of water is expected to reach Hyderabad by November 2015. That means, Godavari River water to Hyderabad in Krishna Basin becomes the first river water interlinking project through Musi River that joints Krishna River downstream of Hyderabad. Chandrababu Naidu [CBN], Chief Minister of AP was in a hurry to show that he is the first person who inaugurated the interlinking of Godavari-Krishna Rivers. For this he used Pattiseema lift irrigation.

DNA [September 19, 2015] presented an article in which it is observed that: "The 174-km-long Polavaram right canal, meant to connect the Polavaram project with the Krishna River was near completion, and what Naidu has done is connect it with another pending small irrigation project, the Pattiseema lift irrigation scheme. The Pattiseema project lifts flood waters from the Godavari when it flows over 15 metres height, and transports it over four kilometres to the Polavaram right canal. The Centre despite granting the project national status [Polavaram] has been reluctant to accelerate it. "

The author presented an article in "telugu360.com" titled "Myths and realities of Godavari – Krishna Inter Linking: An Experts View" dated 15th September 2015. He observed that "I did the interlinking of Godavari River with Krishna River during 1964-69 on many occasions of my travel in train Guntur to Visakhapatnam via Vijayawada to Rajahmundry. I used to fill Krishna water in a bottle at Vijayawada and put that water into Godavari at Rajahmundry and vice versa." AP irrigation minister made a comment on this; but the basics are the same whether he accepts it or not. In both the systems water is lifted intermittently and there is no continuous flow.

Himanshu Thakkar of SANDRP presented a report titled "Godavari Krishna River Linking: Are we celebrating an illegal, unnecessary & misconceived water transfer Project". In this report started with "The national media seems to be celebrating linking of Godavari and Krishna River in Andhra Pradesh on September 16, 2015 as the first major step towards Inter Linking of Rivers in India. An emotional Andhra Pradesh Chief Minister Shri N Chandrababu

Naidu called it *historic* and *Pavitra Sangam* (Holy Confluence). What is the reality?

It states that it is "**Water transfer, not river linking**". Pattiseema Project to lift water from Godavari River from a location about 15 km downstream of the proposed Polavaram dam site and transfer it to an intermediate canal of about 5 km length to already constructed Polavaram Right Bank Canal, which will take it to Krishna River near Ferry/ Ibrahimpatnam village in Krishna district, just upstream of the Prakasham barrage/Vijayawada city (see SANDRP map, **Figure 5.3**). The Pattiseema project involves 30 Nos of Vertical Turbine pumps, each of 8 cumecs (Cubic Meters per Second) capacity. Water was pumped by the existing Tadipudi Lift Irrigation Scheme, a couple of kilometers downstream from Pattiseema. The government has decided to draw 500 cusecs of water from the Tadipudi lift scheme with the help of three pump sets to make Pattiseema scheme operational on an ad hoc basis due to non-completion of the latter's works. SANDPR also raised the question of environmental clearances. The basic issue is not the environmental clearances but the issue is when Polavaram project is taken up as a national project why this? Is it a secret understanding between NDP and TDP governments to serve their vested interests?

In fact, the TDP Government plan was to pump water of Godavari from Pattiseema lift irrigation project in to the Polavaram right canal but the Godavari River water from Tadipudi lift irrigation was pumped in to Polavaram Right Canal. CBN acknowledged this fact but says this is his baby as he laid the foundation stone to Tadipudi lift irrigation project. Unfortunately, NTR laid foundation stones for several projects but none of them started work. CBN also laid second time foundation stones where NTR laid the foundation stones prior to 2004. With this AP lost the flood water use from Krishna basin [8 projects]. All these projects were proposed in Congress time in 70s-80s only. TDP rulers simply laid foundation stones. Under Jalayagnam, Dr. YSR initiated all the projects by giving funds. Now, TDP government, though they abused the Jalayagnam projects, claim Tadipudi lift irrigation project is their own, so they own the interlinking of the rivers whether it is from Tadipudi or Pattiseema lift irrigation project.

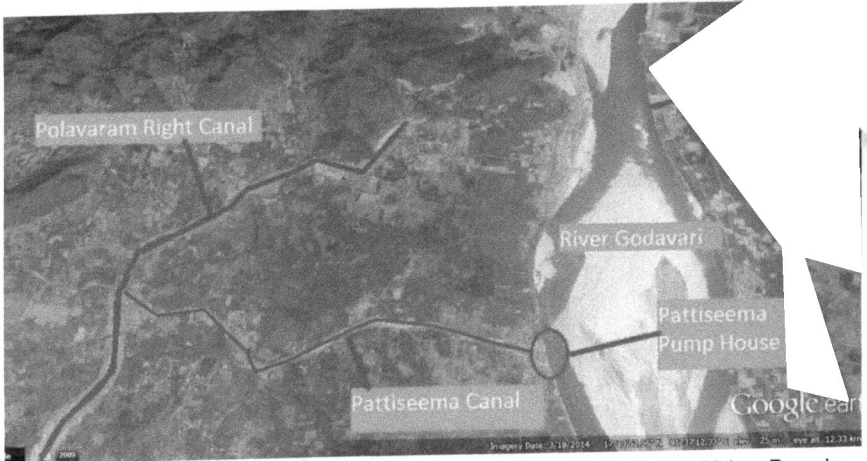

Pattiseema Googla Earth Map (Bhim Singh Rawat, SANDRP) Godavari Krishna Breach map by Bhim Singh Rawant (SANDRP), *Pune, manthan.shripad@gmail.com)*

Figure 5.3: Lift irrigation point of Pattiseema

Have you seen the water that flowed from Tadipudi lift irrigation project built by Dr. YSR government in to Polavaram right canal which was built by Dr. YSR government – it is a muddy water [irrigation minister & TDP leaders have a dip in that water – **Figure 5.4**] pumped in to mud canal [TV channel showed this – water flow cutting the banks and carrying red sandy soil with it and will deposit it in to Krishna River]. CBN inaugurated pumping of Godavari water in to Krishna River at Pattiseema "without a pump" to pump water.

Later using one pump irrigation minister released water behind the water already pumped from Tadipudi. This was carried out in a hurry. This has lead breeching of aqueduct at Tammilleru that connects the Polavaram right canal due to poor quality work similar to Telugu Ganga project during NTR rule. With this, the water that was pumped from Pattiseema joined Tammilleru through which reached Kolleru Lake that connects Bay of Bengal. Even with this quality of work, NDA government assigned the job of preparing DPR for interlinking of few more rivers as presented above!!! A TDP functionary says this will help Rayalaseema. In fact this one pump was secretly brought from Hundri-Neeva lift irrigation Project in Kurnool district. All this was carried out by CBN before he left to Singapore, so as to show the Singapore real estate developers that plenty of water is available for starting the new Capital city. Unfortunately even before he reached ore the pump was stopped as the aqueduct breached. With his Central government has not made any observation on this gross negligence.

Figure 5.4: Water from Tadipudi lift irrigation project

Tanya Singh in the write up on the Pattiseema project says that it is meant to supply water to Rayalaseema. May be the writer has no knowledge on the Rayalaseema projects and Pothireddypadu head regulator which are not yet finished after Dr. YSR death – everybody is afraid that if these are completed Dr. YSR will get the credit. TDP leader including NTR were never interested to take up irrigation projects due to regional and caste based politics. NTR completed Telugu Ganga project with most of the money given by Tamil Nadu Government to take drinking water to Chennai – but on the day of opening the canal breeched submerging vast cropped lands in Rayalaseema and now the same thing happened with Pattiseema wherein within 24 hours of pumping from one motor the aqueduct collapsed with poor foundation.

Tadipudi was built by Dr. YSR government and Polavaram Right Canal was excavated by Dr. YSR government. TDP government simply lifted water from Tadipudi and put in to Right Canal. Therefore the credit, if any, on interlinking as propagated by the media should go to Dr. YSR government and not to TDP government. Tadipudi water is meant for Godavari district and not for Krishna district. Pattiseema lift irrigation project – it is meant to serve the real estate interests in the new capital city only and not to take water to Rayalaseema. Rayalaseema get water only by completing the pending works on projects in Rayalaseema and Polavaram."

Predecessors to Pattiseema: The publicity blitzkrieg by the TDP notwithstanding, Pattiseema is not the first irrigation project that interlinks two rivers in AP. In fact, the much talked about Pattiseema project has three predecessors – two of them dating back to the 19th century. Almost 160 years ago, these two mighty rivers were interlinked through primarily for navigational purposes. The Eluru canals from Godavari and Krishna meet at Malkapuram village near Eluru town to form part of the Kakinada-Pondicherry Buckingham canal. The two Eluru canals also meet the Tammileru rivulet, which drains in to Kolleru Lake. More than 130 years ago, the Tungabhadra River joined the Pennar through the Kurnool-Cuddapah [KC] canal. More recently, the Telugu Ganga Project, which supplies drinking water to Chennai, interconnected the Krishna River with the Pennar River. It also links the Kandaleru River before finally reaching the capital of Tamil Nadu. All these form the part of river linking through natural flow and not through lift irrigation. In fact Polavaram follow the natural flow that comes under real interlinking of "rivers".

5.4 Pros and Cons of Irrigation Systems in India

With growth of the population the water needs for different activities, namely industry, domestic & agriculture are increasing non-linearly under new innovative technologies. We must remember the fact there is a cap on the water availability and in addition this is highly variable. With the urbanization more particularly along the river fronts and chemical input technologies large part of the available water is becoming contaminated and thus reducing the available water. Also, the inefficient management of water is another source of water wastage. This exists in all sectors. For example in the Capital City of Hyderabad around 55% of the drinking water supply is unaccounted in which 40% is coming from the leakages and 60% is coming under pilferage. We rarely control them but in a hurry to bring more water from Krishna and Godavari Rivers at huge cost and diverting land area for those projects.

The major issue that is playing vital role in water utilization is the water diversion to power projects. Majority of the cases this is going as waste. In 2015, both the AP and Telangana governments utilized water from dams even under below the dead storage for power production. Really this water has gone as non-productive. With proper planning wastage would have been reduced to a maximum extent. In the case of Maharashtra the water goes in to Arabian Sea as a waste.

When we are talking of water availability, we forget the fact that this is available only in 75% of the years – dependable condition. Also, under vagaries of monsoon, they are highly variable with year to year, season to season and within the season. Under climate change, the periods of droughts and period of floods drastically change the availability of water. Here the planning is important, particularly in agriculture sector. Unfortunately in ILR DPRs this is not taken in to account and thus may lead to long interstate disputes. When the rains are associated with cyclonic activity the surplus water that enters the Sea/Ocean cannot be taken as true reflection of surplus water as they invariably enter the Sea/Ocean after filling the dams in the downstream. Neither tribunals nor the governments look in to this aspect that creates water war conditions among states and regions within a state.

When we decide a dam, we rarely count the amount of water going to recharge the groundwater except counting the storage on any given day. This will be a major issue with ILR. Unless this is taken in to account, the water balance may not match. When we are talking of submergence due to building of the Dams, we talk of submergence in upstream but we rarely account the fact the submergence associated with floods even without a dam in the downstream areas. The NGO groups fighting against the Dams never cared to look in this direction Similar to our "Waterman". Also, when we build dams or anicuts we forget the problem of pollution in the downstream more particularly associated with urban areas along the rivers.

Sometimes, we waste public money to serve real estate interest on projects. This is the case with Pattiseema lift irrigation. AP government pushed the poor quality Pattiseema lift irrigation project to show to the Singapore realtors that plenty of water is available for the Capital & other corridors construction activities and thus lure them and as well in a hurry to show the world that they are the first to inter-link Godavari River with Krishna River. Here they are trying to link "water" but not "rivers", a fact. A Vastu expert in Hyderabad who got his book released from the President of India, without following the basic rules of traditional Vastu on lying "foundation" built a multi-storied building. On the inauguration day the building collapsed. Also NTR/TDP Telugu Ganga project breached on the inauguration day inundating thousands of acres of cropped area. Now the same TDP government in AP stealthily brought from Hundri-Neeva project one motor and pumped the water from Pattiseema. This water is supposed to join the already pumped water from existing Tadipudi lift irrigation [TDP leaders took bath in this muddy water] water from behind. The aqueduct that connects the Polavaram Right Canal partially collapsed due to poor foundation. If all the pumps were in operation then the entire aqueduct would have been washed away. Government of India should stop such wasteful expenditures and collect that money from the politicians responsible. On 10th October 2015, a report says Chinese experts installed on pump.

5.5 Concluding Remarks

In the case of inter-linking of rivers, particularly in India, the climate change component of natural variability must be taken in to account [**Reddy, 1993**] along with the monsoons periods and cyclone activity periods. Blindly preparing DPRs may result more inter-state and inter-region disputes. They are highly variable from north to the south and east to west. Without understanding the climate, any linking of rivers will become a futile exercise – some of these are explained in **Chapter 3**.

The state governments instead of fighting and making cost escalation in building of projects, they should sit and resolve the issue keeping out the vote bank politics and here the opposition should also join in such negotiations to avoid internal ramblings.

Inter-linking of rivers – gravity flow connects the rivers and life forms in it and this must not be mixed with inter-linking of water – water is lifted only when it is available. Both these systems are in operation in India including in AP.